NATIONAL AUDUBON SOCIETY POCKET GUIDE

NATIONAL AUDUBON SOCIETY

The mission of the NATIONAL AUDUBON SOCIETY *is to conserve and restore natural ecosystems, focusing on birds and other wildlife for the benefit of humanity and the Earth's biological diversity.*

We have 500,000 members and an extensive chapter network, plus a staff of scientists, lobbyists, lawyers, policy analysts, and educators. Through our sanctuaries we manage 150,000 acres of critical habitat.

Our award-winning *Audubon* magazine, sent to all members, carries outstanding articles and color photography on wildlife, nature, and the environment. We also publish *Audubon Field Notes,* a journal reporting seasonal bird sightings continent-wide; *Audubon Activist,* a newsjournal, and *Audubon Adventures,* a newsletter reaching 600,000 elementary school students. Our *World of Audubon* television shows air on TBS and public television.

For information about how you can become a member, please write or call the Membership Department at:

NATIONAL AUDUBON SOCIETY
700 Broadway
New York, New York 10003
(212) 979-3000

CONSTELLATIONS OF THE NORTHERN SKY

Text by Dr. Gary Mechler and Dr. Mark Chartrand

Astronomical Charts by Wil Tirion

Alfred A. Knopf, New York

This is a Borzoi Book
Published by Alfred A. Knopf, Inc.

 Published in the United States by Alfred A. Knopf, Inc., New York, and simultaneously in Canada by Random House of Canada Limited, Toronto. Distributed by Random House, Inc., New York.

Prepared and produced by Chanticleer Press, Inc., New York.
Typeset by Chanticleer Press, Inc., New York.
Printed and bound by Tien Wah Press, Singapore.

Published March 1995
First Printing

Library of Congress Catalog Card Number: 94-34032
ISBN: 0-679-77998-1

Contents

How to Use This Guide

Human beings have always looked to the skies: as clock and calendar, to divine the future, for spiritual guidance. Even today, astronomers study the night sky for the key to the mysteries of the universe, as they literally look back in time to when it all began. By learning your way around the sky, you can identify constellations that our ancestors knew and even search out distant galaxies and clusters of stars that may support worlds like our own.

Coverage

This guide covers the Northern Hemisphere skies season by season, with closeup looks at 54 constellations visible from northern latitudes, both the familiar and the less well known.

Organization

This easy-to-use pocket guide is divided into four parts: introductory essays and drawings, illustrated tours of the sky, illustrated descriptions of the constellations, and appendices.

Introduction

The essay "What Is a Constellation?" examines both the popular and the scientific concept of the constellation. "The Celestial Sphere" discusses the fixed backdrop of stars that we see as Earth makes its annual journey along its orbit. The section "Stars," with essays on star formation, star types, colors, sizes, and brightness, and multiple and variable stars, explains how these distant objects evolve, how they are classified, and how they relate to one another. "Deep-Sky Objects" defines objects that exist beyond our solar system, such as star clusters, nebulae, and galaxies. Finally, "Observing the Sky" offers stargazing tips and explains how to use the color charts of the northern skies and the individual constellations. Following the Introduction are a table of the Greek alphabet, to which you may want to refer when learning to identify stars with Greek-letter names, and a key to the symbols used in the charts.

The Sky Tours

This section contains 16 color charts of the northern skies, showing the stars and constellations visible in the northwestern, northeastern, southwestern, and southeastern quadrants of the sky in each season. The accompanying text provides a quick tour of that part of the sky, pointing out objects of interest.

The Constellations

Fifty-four constellations that can be viewed from northern latitudes are presented in this section. A color chart of each constellation shows its boundaries, its characteristic shape, and its stars and deep-sky objects. Facing each chart is information about the constellation and the stars and deep-sky objects most accessible to the amateur observer.

The Appendices

The table of the southern constellations gives the names and locations of the 34 constellations visible in the Southern Hemisphere that we don't cover here. "Systems of Measurement" explains celestial coordinate systems, used to locate objects in the sky. For quick reference, turn to the tables "Luminosity Classes" and "Spectral Types" to find out a star's color and temperature and its phase in its life cycle, or scan the list of the brightest stars in the sky. The glossary explains common terms used by astronomers and observers. The constellations, named stars, and deep-sky objects discussed are listed in the index.

Learning about the constellations and the stars and other objects within them will enrich your understanding of our place in the universe—and increase your appreciation of our ancestors of the ancient world, who knew how to read the skies.

What Is a Constellation?

In ancient times human beings imagined they could see the outlines of figures in the stars of the night sky. These figures, given the shapes of mythological heroes, creatures, and objects, came to have their own legends and were even thought to influence events on Earth. This remains the popular concept of what a constellation is. In modern astronomy, however, the term constellation refers to a particular region of the sky, often enclosing the "figure" that may first have been noticed millennia ago, but also including the surrounding area. These designated regions of the celestial sphere are like states or countries on a map; every portion of the sky is now said to belong within a particular constellation. The borders between the constellations are linear, but they can be quite irregular, given the sizes and shapes of the star groups they enclose.

The concept of a constellation is simply a convenience, indicating a direction of the sky toward which we can look to find a specific object. From our point of view all the stars within a constellation seem to be physically related. We cannot, with the naked eye, distinguish among different depths in space, and so the stars that we can see all seem to be at the same distance and on the same plane.

In reality, most of the individual stars are quite far apart; those visible to the naked eye are typically separated by hundreds, even thousands of light-years.

Ancient and Modern Constellations

Among the ancient civilizations that first named constellations were the Babylonian, Indian, Greek, Roman, Chinese, and Native American. These people dwelt in the Northern Hemisphere and were therefore able to name only groups of stars visible in northern latitudes, as the far southern constellations were not visible to them. The second-century Greco-Egyptian astronomer Ptolemy cataloged more than 1,000 stars and 48 constellations in his work the *Almagest.* These constellations that have been known since antiquity are called ancient constellations.

It was not until European navigators began exploring south of the equator in the 16th century that new cataloging was done and that Southern Hemisphere constellations began to be defined for the Western world. In the early 17th century, Johann Bayer named 12 Southern Hemisphere constellations; his countryman and contemporary Jakob Bartsch named three others. Johannes Hevelius named seven more in 1687; and

Nicolas-Louis de Lacaille, after a trip to the southern tip of Africa, named 14 more and cataloged about 10,000 stars between 1750 and 1754. These new constellations are known as modern constellations.

Generally the ancient constellations are named for the "figure" drawn with their stars. Orion and Leo are two that look something like the beings for which they were named. Many of the modern constellations were named for friends, patrons, and inventions, such as the microscope and the telescope. The shapes (i.e., the lines connecting the stars) of specific constellations are somewhat arbitrary and vary somewhat on different sky maps. A few constellations contain smaller, identifiable shapes within them, known as *asterisms.* The Big Dipper is an asterism within the constellation Ursa Major.

Before 1930 anyone could name any part of the sky anything, and there were no widely accepted constellation boundaries. To eliminate the confusion, in 1930 astronomers worldwide decided to specify the names (in Latin) and boundaries of 88 constellations. These are the constellations universally recognized today.

The Celestial Sphere

Ancient skywatchers perceived the sky as a sphere with a fixed background (the stars) against which the solar system bodies move. Many of their findings and concepts are still used by stargazers today.

The Zodiac and the Ecliptic

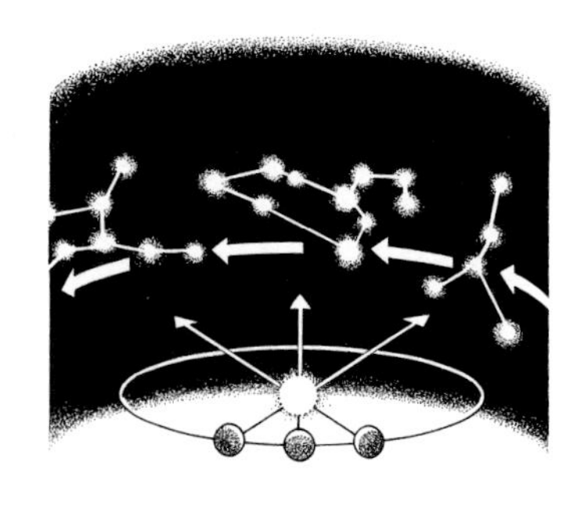

One of the first areas of the night sky you may want to observe is the zodiac and its central line, the ecliptic, the band of the sky through which, from our point of view on Earth, the Sun, Moon, and planets move. As the Earth makes its annual journey along its orbit, the Sun seems to trace a path through the sky against the backdrop of stars. That apparent path is known as the ecliptic, and the fixed backdrop of stars, extending about 8° north and south of it, is the zodiac. Because all the planets (except Pluto) orbit in more or less the same plane as Earth, they too move through the zodiacal band.

In about the fifth century B.C., Babylonian astrologers (and perhaps Greek skywatchers as well) conceptualized the ecliptic and divided it into 12 segments, each about 30° wide and marked by a constellation. All but one of these constellations (Libra, the Scales) represent a real or imaginary creature, hence the name *zodiac*, which means "circle of animals."

Precession

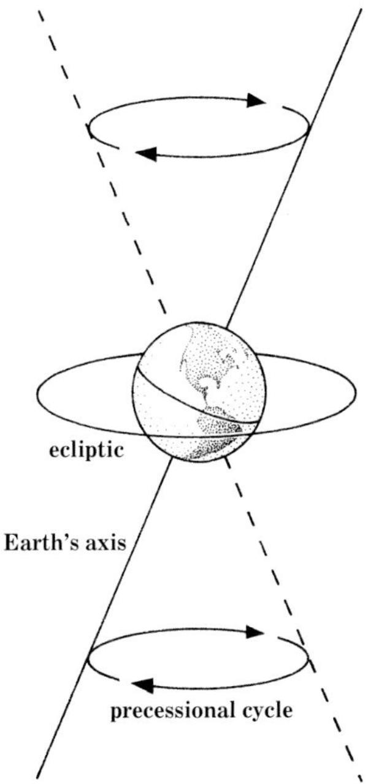

Our ancient forebears also knew, by the second century B.C., about a process called precession, a slow gyration of the Earth's axis (it takes the axis about 26,000 years to complete one gyration). Because of the precessional change in the direction of the axis, the Earth is now positioned slightly differently relative to other celestial bodies than it was in antiquity. One of the effects of this is that the Sun now enters a given constellation about a month later than it did two thousand years ago. Then the Sun entered Aries at the time of the vernal equinox (the beginning of spring in the Northern Hemisphere). Today the Sun is in Pisces at the time of the vernal equinox and doesn't enter Aries until a month later.

Another effect of precession is that the polestars change over time. The polestars mark the north and south celestial poles, extensions of the terrestrial north and south poles into space, as determined by the direction of the Earth's axis. As the axis gyrates with precession, the celestial poles change. The star Thuban, the α star in Draco, occupied the position of the north celestial pole 4,000 years ago. Today Polaris, the α star in Ursa Minor, is the North Star; no star marks the south celestial pole.

Stars

Dozens of prominent stars, such as Sirius and Polaris, possess common names that date back hundreds or even thousands of years, often to Greek, Roman, and Arabic sources. In 1603, in his famous star atlas *Uranometria*, German astronomer Johann Bayer instituted a system of assigning Greek letters to stars. A Bayer designation consists of a lowercase Greek letter followed by the genitive (possessive) form or the abbreviation of the name of the constellation. Sirius, for example, is the alpha star of Canis Major, or α Canis Majoris, or α CMa. The letters usually are assigned to stars in the order of their brightness within a given constellation, with the brightest star usually called α (alpha), the second-brightest β (beta), the third γ (gamma), and so on, according to the sequence of letters in the Greek alphabet (see the chart of the Greek alphabet, p. 31). Most bright stars have a Greek-letter designation, but a few have labels from another system.

Bayer's system was introduced before the constellations were formally codified in 1930; therefore, some stars near constellation boundaries are no longer in their original constellations or have been renamed.

Stellar Evolution

Stars are composed mainly of hydrogen gas. Their light comes from the energy produced at their cores by nuclear fusion. Stars form within clouds of gas and dust in space when random swirling motions, collisions between clouds, or the explosion of a nearby star cause the cloud to contract until it reaches a stage called critical density, at which point the mutual gravity of all the atoms becomes strong enough to continue pulling the cloud together. As the cloud shrinks it compresses and heats up, ultimately reaching a temperature of several million degrees at its center. The high pressure and temperatures permit nuclear reactions to occur, specifically, the conversion of hydrogen to helium and energy. Stars in this hydrogen-using stage are known as *main sequence stars.* Our Sun has been in this stage for about 4.5 billion years and has another 5 billion years to go before using up its hydrogen fuel. At later stages in a star's life cycle, helium fuses into carbon, oxygen, and other elements, which in turn may fuse into heavier elements. This continues the life of the star beyond the depletion of hydrogen and can produce such heavy elements as iron and uranium.

When a star has exhausted the hydrogen fuel at its center it begins to expand and its surface gets cooler. The star reaches enormous proportions and becomes a *red giant* or *supergiant,* depending on its mass. After swelling to many times its former size and using up its store of helium, the red giant sheds its outer envelope, which becomes a planetary nebula (see below). Meanwhile, the star's interior begins to shrink, its surface heats up, becoming white hot, and it becomes a *white dwarf,* an extremely dense star. A teaspoonful of the matter of such a star would weigh many tons. Very massive stars become supergiants and may grow as large around as the orbit circumscribed by the planet Jupiter. Late in their lives such massive stars become unstable and thus variable in light output, and a very few become *supernovas,* exploding spectacularly. Supernovas blast off most of their material, leaving behind supernova remnants (see below) and a tiny but incredibly compressed core called a *neutron star* or *pulsar.* This small star usually rotates rapidly, sending out beams of light and radio waves. Some neutron stars may become *black holes,* regions so dense that even light cannot escape their intense gravitational fields.

Star Brightness

The naked eye is capable of detecting three things about a pointlike light, such as a star: brightness, color, and direction. The brightness of a star as seen from Earth is called its apparent magnitude. About 2,000 years ago the Greek astronomer Hipparchus designated the brightest stars in the sky "1st magnitude" and those just barely visible to the unaided eye "6th magnitude." In the 19th century British astronomer Norman Pogson quantified the scale more precisely so that each step in magnitude equals a factor in brightness of 2.512 times. A star of 1st magnitude is therefore almost exactly 100 times (2.512^5) brighter than a star of 6th magnitude. The brightest objects in the sky are brighter than 1st magnitude; their apparent magnitudes are given as negative numbers. The brightest star in the sky, Sirius (α Canis Majoris) has a magnitude of -1.46, and Venus, at its brightest, about -4. The Sun's apparent magnitude is about -27. The apparent magnitude of an object depends on both its intrinsic brightness and its distance from Earth. An intrinsically faint star, if close to Earth, can appear brighter than an intrinsically bright star that is much more distant.

Star Color and Size

A star's temperature can be deduced from the star's color—that is, from the spectrum of the light it emits. Red stars, such as Antares (α Scorpii), are coolest, with a surface temperature of just 3,000° Kelvin; yellow stars, such as the Sun, are medium hot, about 5,800°K on the surface; white stars, such as Procyon (α Canis Minoris), reach 7,500°K; the very hottest blue stars, such as the three stars in Orion's Belt, are more than 50,000°K. (To convert the Kelvin, or absolute, scale to Fahrenheit, use this formula: $°F = [1.8 \times °K] - 459$.)

Astronomers have devised a system of classification, called spectral type, that groups stars by the strengths and positions of absorption lines in their spectra. These absorption lines are a function of temperature. From hottest to coolest (that is, bluest to reddest) the seven major spectral types are O, B, A, F, G, K, M. (A mnemonic device for remembering these is "Oh, Be A Fine Girl/Guy, Kiss Me.") Although all stars have some color corresponding to their surface temperature, most stars appear white to the naked eye because the eye is not sensitive to color at low light levels. We can detect the surface colors only of stars of about 1st magnitude or brighter. Each spectral type is divided into a number of subtypes, usually 0 through 9, hotter to cooler.

A star's brightness is also a function of its size. While hotter stars tend to be intrinsically brighter, and cooler stars tend to be dimmer, some cool giant and supergiant stars are very bright because they are so large. Another classification system, called luminosity class, groups stars according to size. Main sequence stars are luminosity class V; supergiants, depending on brightness, may be luminosity class Ia, Iab, or Ib; and white dwarfs are luminosity class VII. The full classification of a star includes its spectral type and its luminosity class. Our yellow Sun, a main sequence star, is type G2 V.

Multiple and Variable Stars

Most stars have at least one stellar companion; they are called double (or binary), triple, or multiple star systems, depending on the number of stars. Multiple stars are often interesting objects for viewing in binoculars or small telescopes.

Physical multiple star systems are gravitationally bound, and the stars within them orbit one another. *Optical multiple stars* are unrelated stars, often quite far apart, that appear to be close together along our line of sight. Some physical multiple star systems, called *spectroscopic binaries* or *multiples,* are so close to one another that they cannot be resolved into individual stars. The presence of more than one star can only be inferred from the spectrum of light emitted by the system. Another type of double star system is the *eclipsing binary.* One star in such a system periodically eclipses the other as it passes in front of it during their mutual orbit, causing the star to dim for a few hours at a time.

Despite the variations in its magnitude, the eclipsing binary is not a true variable star. Variables are single stars that vary in actual light output. The main categories are *pulsating variables,* giant or supergiant stars that expand and contract, varying in temperature and magnitude (there are several types of pulsating variables); *eruptive variables,* stars that experience explosions, including novas and supernovas; and *rotating variables,* stars with a dark, cool spot on the surface (similar to a sunspot), which causes a decrease in brightness when the cool spot rotates into our view.

Deep-Sky Objects

Deep-sky objects are basically groups of stars, such as star clusters and galaxies, or nonstellar objects, such as nebulae, that exist beyond our solar system. These objects have been cataloged by astronomers, and hundreds of special lists exist. The most common listings are the Messier (M) list and the New General Catalog (NGC) and its supplement, the Index Catalog (IC).

In the late 1700s French astronomer Charles Messier, known as the "comet ferret" by associates, was often confused by "fuzzy objects" that we now recognize as star clusters, nebulae, and galaxies. He compiled a list of these "objects to avoid" that frustrated his efforts to find comets. Later astronomers have slightly modified and extended Messier's original list, and these hundred-odd "Messier objects" are today among the most popular targets for amateur stargazers. The New General Catalog and Index Catalog of deep-sky objects were compiled in the late 19th century and have been revised since. Many deep-sky objects have both Messier and NGC (or IC) numbers.

Star Clusters

Star clusters are groups of gravitationally bound stars similar to one another in age and composition that formed from the same interstellar cloud at about the same time.

Open clusters (also called galactic clusters) are loose groupings of dozens to a few thousand relatively young stars, the brightest of which are hot blue stars. They often contain much interstellar gas, inside which new stars may be forming. The loosest open clusters are called *associations.* The Pleiades, in the constellation Taurus, is a relatively compact, young open cluster.

Globular clusters are roughly spheroidal groups of up to several million stars that formed early in the evolution of their galaxy. The stars are old red giants and supergiants, and there is little or no interstellar gas among them. Most globulars are very far from Earth. The brightest appear as fuzzy stars to the naked eye.

Nebulae

Nebulae are clouds of gas (mostly hydrogen) and dust (mostly carbon and silicon) in space. The brighter nebulae are quite beautiful sights in small telescopes. The types of nebulae are: *absorption nebulae,* dark clouds in which the dust absorbs or scatters the light of more distant stars; *reflection nebulae,* illuminated by the glow of a nearby star reflecting off the dust; *emission nebulae,* brightly lit by fluorescence of gases caused by irradiation from nearby stars; *planetary nebulae,* thin, sometimes ring-shaped shells of gas thrown off by an evolving, gently erupting star at the end of its red-giant phase; and *supernova remnants,* huge, irregular, wispy shells of expanding gases that result from a supernova explosion. Absorption, emission, and reflection nebulae are also called *diffuse nebulae.*

Galaxies

Galaxies are aggregates of gas, dust, and millions or billions of stars held together by mutual gravitational forces. Galaxies occur in clusters, and clusters of galaxies occur in superclusters. Our Milky Way Galaxy belongs to the Local Group galaxy cluster and the Virgo Supercluster. *Elliptical galaxies* (type E) range in shape from almost spherical to highly elongated. They contain little or no gas and dust and tend to consist almost entirely of old stars. *Spiral galaxies* (type S) are those with a central structure from which curving arms extend (the Milky Way is a spiral). A *barred spiral galaxy* (type SB) is a type of spiral with a bar of stars and interstellar matter running through its nucleus. *Irregular galaxies* (type Irr) have no regular shape or only a hint of one and are usually rather small. Galaxies that don't fit neatly into these categories are deemed *peculiar galaxies.*

Observing the Sky

What you see in the sky is largely determined by the time of year, the time of day or night, your location on Earth, and the condition of the atmosphere. Once you are outdoors it will take several minutes for your eyes to adapt to the darkness. To maintain your adaptation do not use a white light for reading the sky charts or checking instrument settings. Use a small flashlight covered with a red filter or with several layers of red cellophane.

Before you step outside to view the night sky, take a few minutes to decide what it is you want to see and how to find it. The seasonal sky charts in this guide show the skies for each season, locating the constellations and bright stars visible in the evening at that time of year. The individual constellation charts focus on particular constellations visible in the northern skies, locating and describing the bright stars and interesting deep-sky objects accessible to the amateur observer.

The trick to finding your way around the sky is to look first for the most distinctive constellations, then use them as signposts and pointers to other constellations. Several of the best stellar guides are Orion, the face of Taurus and the nearby Pleiades star cluster, the Big

Dipper (in Ursa Major), the W shape of Cassiopeia, and Scorpius. Somewhat less bright but still quite noticeable are the Great Square of Pegasus, Cygnus, Leo, and the trapezium of Corvus.

If you wish to search for deep-sky objects, note that constellations along the plane of the Milky Way tend to be rich in open star clusters and nebulae. You will have to look away from the plane, out into deep space beyond our galaxy, to search out other galaxies and globular clusters.

The Sky Tours

The first part of the color-plate section, the Sky Tours, includes 16 seasonal sky charts, each accompanied by a brief narrative tour of the part of the sky pictured. Because it is impossible to accurately portray the entire dome of the sky on a flat page, for each season there are four charts oriented to the intercardinal directions: northwest, northeast, southwest, and southeast. The compass illustration indicates which quadrant of the sky is featured. Simply hold the chart and face the direction pictured on that chart. These charts plot the constellations and major stars. The wavy, pale blue areas represent the band of the Milky Way. The zenith, the point directly overhead, is indicated on each chart.

The seasonal sky charts are drawn to show the sky from latitude 35° N at the times and dates specified in the text accompanying each chart. They can, however, be used over a wider range of mid-northern latitudes and times. If you are observing from a latitude north of 35°, remember that everything in the northwestern and northeastern quadrants of the sky will be a bit higher in the sky, and all the objects in the southeastern and southwestern quadrants will be a bit lower in the sky. Keep in mind that a two-hour difference in time of night is the equivalent of a one-month difference in date. Thus the positions of objects in the sky at 9:00 P.M. on January 15 will be the same as at 11:00 P.M. on December 15 and at 7:00 P.M. on February 15. By observing very late at night or early in the morning you can see the early-evening stars of other seasons.

These seasonal charts illustrate only stars and constellations, as the positions of the solar system objects change continually. If you wish to find the planets among the stars, you may find the locations of planets (what constellations they can be found in) listed on the weather page of your local newspaper.

The Sky Tours, like any guided tour, will help you find your way around the sky. Once you have located an area that interests you, you may then turn to the individual constellation chart to study it in more depth.

The Constellation Charts

Following the 16 seasonal sky charts are charts of the 54 constellations most easily visible from the Northern Hemisphere. We have included constellations as far as 30° to 35° south of the celestial equator (declination −30° to −35°). The more southerly constellations, some of which can be glimpsed from southern areas of North America, are listed in a table in the appendices.

The constellation charts are presented in alphabetical order. Each chart shows the boundaries of the constellation and outlines its characteristic shape. Stars are labeled with their Greek-letter designations, and deep-sky objects of special interest are labeled with their Messier, NGC, or IC numbers. (Messier numbers are preceded by the letter M, IC numbers with "I", and NGC numbers have no prefix.) Preceding the color-plate section is a table of the Greek alphabet and a key to the charts, indicating the symbols plotted on the charts for stars of different magnitudes and for various deep-sky objects. Celestial coordinates—declination (degrees) and right ascension (hours)—are also plotted on the charts. (See the appendix "Systems of Measurement" for an explanation of celestial coordinate systems.) Some charts have a black dotted line, indicating the ecliptic, or a wavy, pale blue area, representing the band of the Milky Way.

Note that on the constellation charts east and west appear to be reversed. When you look at a land map, you are looking from above. When you look up into the sky, however, you are looking from below, and east is to your left and west is to your right.

The Constellation Descriptions

Opposite each constellation chart is the accompanying text. Two constellations have received slightly different treatment: Hydra begins with two pages of text, followed by a two-page chart, due to its great size. Half of Serpens, the only noncontiguous constellation, called Serpens Cauda, is plotted on the Scutum chart, while Serpens Caput has its own chart. Both are described as one constellation on the page opposite the chart for Serpens Caput, which immediately follows Scutum.

For each constellation we provide the Latin name formally adopted by the International Astronomical Union in 1930, followed by the abbreviation, a standardized three-letter form. "On meridian" gives the date on which the constellation culminates (i.e., is on the observer's celestial meridian, the imaginary north-south line running from the celestial poles and passing through the observer's zenith) at 9:00 P.M. local standard time (10:00 P.M. daylight saving time). This is followed by the constellation's size, the measure of its area as defined by its official boundaries, as viewed on the sky. In the margin is an illustration of the form or figure the constellation's stars usually represent.

The description of the constellation begins with general comments about the constellation, including its history, mythology, and any particularly outstanding celestial object within its boundaries. The second paragraph describes notable stars, usually the brightest, in the constellation. This section includes the star's Greek-letter designation, common name (if any), spectral type and luminosity class, apparent magnitude, distance from Earth in light-years (ly), and other relevant information, such as whether it is variable or part of a multiple star system. The final section describes notable deep-sky objects—star clusters, nebulae, galaxies—found within the constellation's boundaries. Information on their magnitude, distance from Earth in light-years, location relative to other objects, and apparent size (explained in the appendix "Systems of Measurement") may also be provided. Most deep-sky objects are assigned either a Messier number (which starts with an M), a New General Catalog number (NGC), or an Index Catalog number (IC).

For the first-time stargazer, attempting to relate a small, two-dimensional map to the majestic dome of the sky can be frustrating. It will take time to absorb and remember constellation shapes, their relationships with other star groups, and names of stars and other objects. But the reward for your troubles is great: You will have begun to make the universe your own. So grab your jacket and a Thermos of hot tea or coffee, step outside, and get comfortable. Good observing!

KEY FOR THE SKY CHARTS

MAGNITUDES -1 0 1 2 3 4 5

Double stars **Variable stars** **Open clusters**

Globular clusters **Planetary nebulae**

Diffuse nebulae **Galaxies**

THE GREEK ALPHABET

α	Alpha	ι	Iota	ρ	Rho
β	Beta	κ	Kappa	σ	Sigma
γ	Gamma	λ	Lambda	τ	Tau
δ	Delta	μ	Mu	υ	Upsilon
ϵ	Epsilon	ν	Nu	φ	Phi
ζ	Zeta	ξ	Xi	χ	Chi
η	Eta	ο	Omicron	ψ	Psi
ϑ	Theta	π	Pi	ω	Omega

peja.
PERSEUS.
Andromeda.
Via

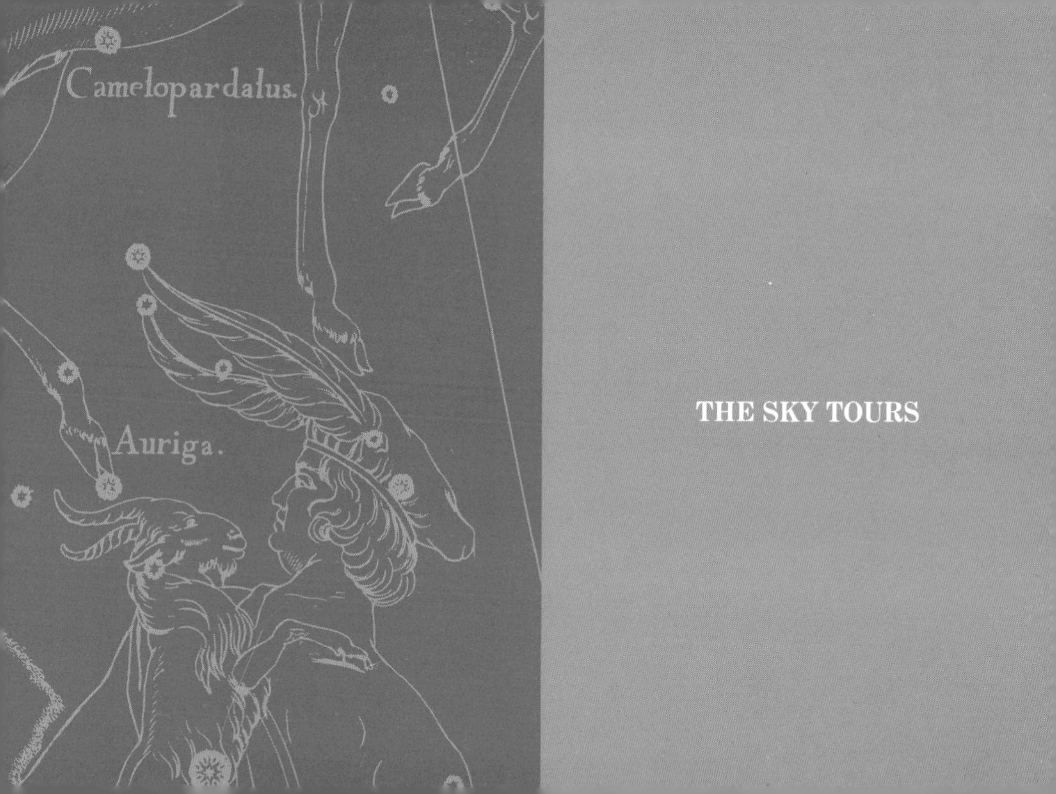

THE SKY TOURS

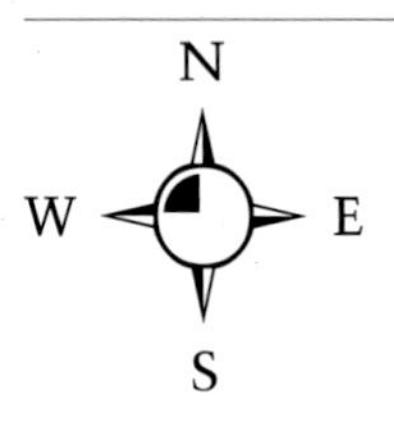

THE WINTER SKY: Northwest

Chart shows: December 15 sky, 11 P.M.
January 15 sky, 9 P.M.
February 15 sky, 7 P.M.

Low in the northwestern sky, the last of the late-summer and fall constellations are just about to set. Ursa Minor, the Little Bear, hangs almost vertically from Polaris (α UMi), the North Star, which marks the tip of its tail or, alternatively, the end of the handle of the Little Dipper. Deneb, the α star in Cygnus, is just above the horizon. The Great Square of Pegasus is almost due west, with the stars of Andromeda running upward from its uppermost corner. Between Pegasus and the North Star are Cepheus, the King, and Cassiopeia, the Queen. Above Andromeda, near the zenith, lies Perseus, the Hero. The Milky Way runs up through Cygnus, Cassiopeia, Perseus, and Auriga, and down into the eastern sky and the winter constellations.

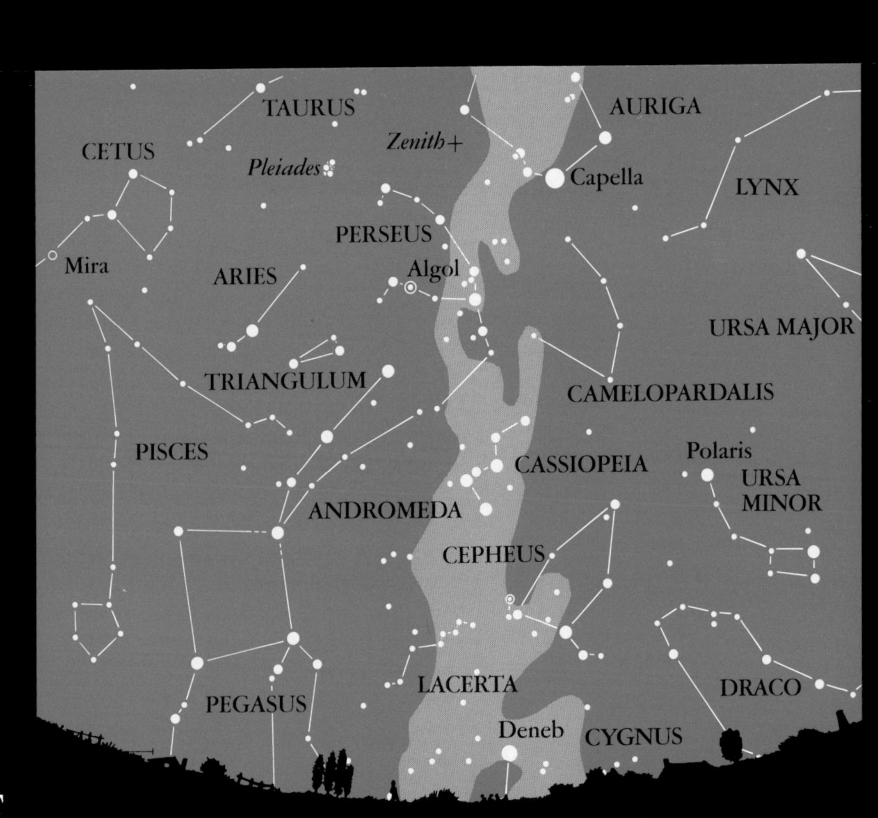
TAURUS
AURIGA
CETUS
Zenith +
Pleiades
Capella
LYNX
PERSEUS
Mira
Algol
ARIES
URSA MAJOR
TRIANGULUM
CAMELOPARDALIS
PISCES
Polaris
CASSIOPEIA
URSA
MINOR
ANDROMEDA
CEPHEUS
LACERTA
DRACO
PEGASUS
Deneb
CYGNUS

WEST

NORTH

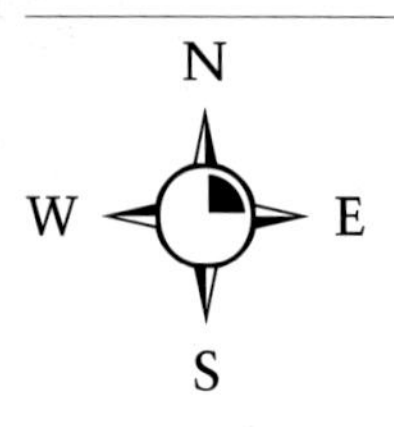

THE WINTER SKY: Northeast

Chart shows: December 15 sky, 11 P.M.
January 15 sky, 9 P.M.
February 15 sky, 7 P.M.

Low in the northeastern sky is the Big Dipper (an asterism in the constellation Ursa Major), standing upright on its handle. The two stars that form one side of the Dipper's bowl (α and β UMa) are known as the Pointers and show the way to Polaris (α UMi), the North Star, in the Little Dipper. Polaris marks the north celestial pole, the extension of the North Pole, and Earth's axis, into space. Leo, the Lion, and its bright α star, Regulus, are just rising to the east. Above Leo lies the faint zodiacal constellation Cancer, the Crab. Higher up are Procyon (α CMi), sometimes called the Little Dog Star, in Canis Minor, and Castor (α Gem) and Pollux (β Gem), the Twins, in Gemini.

AND
PERSEUS
Algol
TAURUS
Zenith+
ORION
Betelgeuse
Capella
CASSIOPEIA
AURIGA
GEMINI
MONOCEROS
CAMELOPARDALIS
CANIS MINOR
Castor
Procyon
LYNX
Pollux
CEPHEUS
Polaris
URSA MINOR
CANCER
URSA MAJOR
HYDRA
LEO
Regulus
LEO MINOR
DRACO
SEX
NORTH
EAST

N

W E

S

THE WINTER SKY: Southeast

Chart shows: December 15 sky, 11 P.M.
January 15 sky, 9 P.M.
February 15 sky, 7 P.M.

The southeastern sky is dominated by Orion, the Hunter, perhaps the most famous constellation. The three equally bright, evenly spaced white stars in a straight line mark his belt. A line drawn along the belt and extended to the west points roughly to the reddish star Aldebaran (α Tau), the eye of Taurus, the Bull, and farther still to the Pleiades star cluster. A line extended from the belt to the east points toward the brightest star in the sky, Sirius, the Dog Star (α CMa). Marking Orion's eastern shoulder is the red supergiant Betelgeuse (α Ori). The bright star marking his western knee is the bluish-white giant Rigel (β Ori). A line drawn from Rigel through Orion's belt to Betelgeuse and extended northward points just north of Gemini, the Twins, and its brightest stars, Castor (α Gem) and Pollux (β Gem). Between Gemini and Sirius lies the bright star Procyon, the α star in Canis Minor, the Little Dog. Above Orion in the east is Capella, the α star in Auriga, the Charioteer.

PERSEUS
Capella
Pleiades
+Zenith
URSA MAJOR
AURIGA
LYNX
Aldebaran
TAURUS
CETUS
GEMINI
Castor
ORION
Pollux
Betelgeuse
Rigel
ERIDANUS
CANCER
CANIS MINOR
Procyon
LEO
MONOCEROS
LEPUS
Sirius
CAELUM
Regulus
CANIS MAJOR
HYDRA
HOR
COLUMBA
SEXTANS
Adhara
PUPPIS

EAST

SOUTH

THE WINTER SKY: Southwest

Chart shows: December 15 sky, 11 P.M.
January 15 sky, 9 P.M.
February 15 sky, 7 P.M.

There are few bright stars in the southwest at this time. The dim but large constellations of Pisces, the Fish; Cetus, the Sea Monster; and Eridanus, the River, span most of the western and southwestern sky. Not surprisingly, the ancients referred to this area of the sky, filled with sea creatures, as "the Water" or "the Sea." High in the south-southwest is Taurus, the Bull, the bright orange star Aldebaran (α Tau) marking his eye; the V-shaped Hyades star cluster marks his face. High in the southwest is the attractive Pleiades star cluster, also in Taurus. Almost overhead is one of the brightest stars in the sky, Capella, the α star in Auriga, the Charioteer.

AURIGA
PERSEUS
CASSIOPEIA
GEMINI
MON
+Zenith
Algol
Betelgeuse
Aldebaran
ORION
Pleiades
TRIANGULUM
ANDROMEDA
Wil Tirion
ARIES
Rigel
LEPUS
PISCES
Mira
PEGASUS
COLUMBA
ERIDANUS
CETUS
CAELUM
FORNAX
HOR
SCULPTOR

OUTH

WEST

THE SPRING SKY: Northwest

Chart shows: March 15 sky, 11 P.M.
April 15 sky, 9 P.M.; 10 P.M. DST
May 15 sky, 7 P.M.

Say good-bye to Orion and the stars of winter. During this season you can see them in the evening low in the northwest until mid-April. Taurus, and its bright orange-red α star, Aldebaran, are very low in the northwest. Capella, the α star in Auriga, shines brightly above Taurus. The fleur-de-lis shape of Perseus is below Capella, and the W of Cassiopeia is low in the northern sky, just above the horizon. Castor (α Gem), Pollux (β Gem), and Procyon (α CMi) are above and to the west of Capella. The faint stars of Lynx, so-named because a viewer needs the sharp eyes of that nocturnal cat to see them, extend above Capella. The Milky Way is intrinsically faint here. It is brightest at Cassiopeia, where it reaches the horizon and is so difficult to see because of atmospheric interference and light pollution.

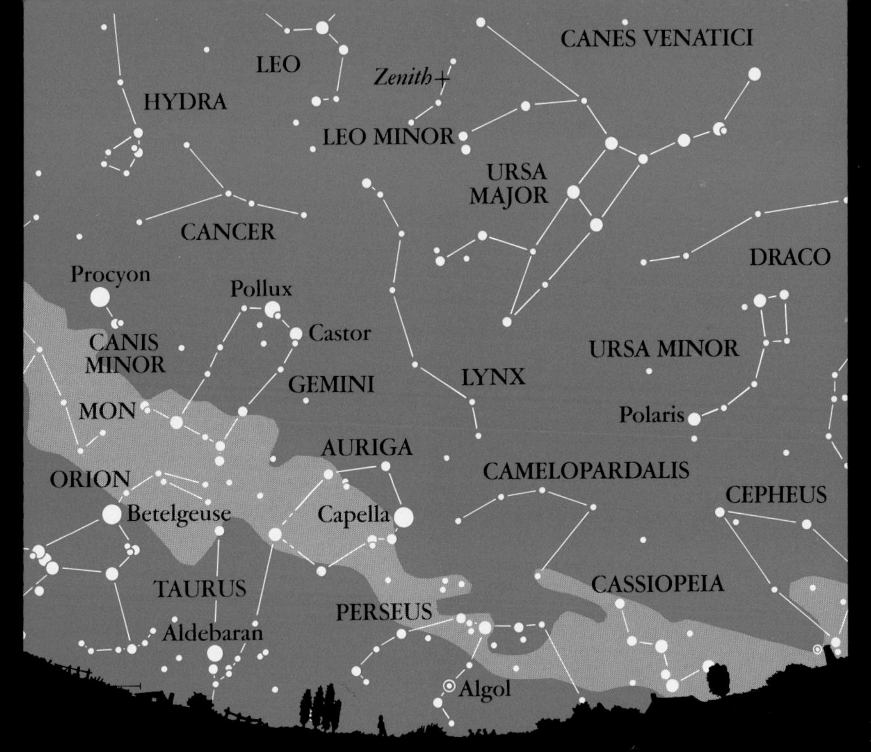
CANES VENATICI
LEO
Zenith+
HYDRA
LEO MINOR
URSA MAJOR
CANCER
DRACO
Procyon
Pollux
Castor
CANIS MINOR
URSA MINOR
LYNX
GEMINI
MON
Polaris
AURIGA
CAMELOPARDALIS
ORION
CEPHEUS
Betelgeuse
Capella
TAURUS
CASSIOPEIA
PERSEUS
Aldebaran
Algol

VEST

NORTH

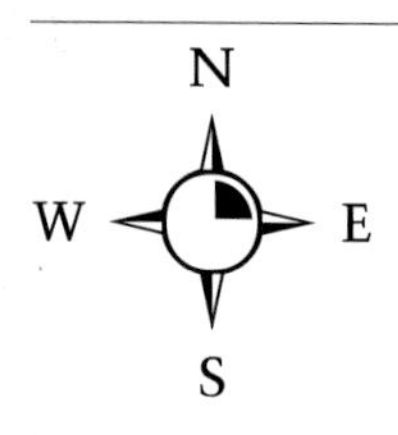

THE SPRING SKY: Northeast

Chart shows: March 15 sky, 11 P.M.
April 15 sky, 9 P.M.; 10 P.M. DST
May 15 sky, 7 P.M.; 8 P.M. DST

High in the northeast is Ursa Major, the Great Bear, whose hindquarters most people recognize as the Big Dipper. Two stars in its bowl, α and β Ursae Majoris, are called the Pointers because a line drawn through them and extended northward points to Polaris (α UMi), the North Star. The same line, extended away from Polaris, points to Leo, high in the southeast. If you follow the arc of the Dipper's curved handle eastward, you will "arc to Arcturus," the bright orange star (α Boo) midway up in the eastern sky that marks the base of the kite-shaped figure of Boötes, the Herdsman, now lying almost horizontal. Continuing the line from Arcturus you will "speed to Spica," the α star in Virgo, in the southeast. Below Arcturus is the semicircular Corona Borealis, and below that is the butterfly or H-shape of Hercules. Just rising in the northeast is the sparkling white star Vega, the α star in Lyra, first of the summer stars.

LYNX
LEO MINOR
Zenith
LEO
VIRGO
URSA MAJOR
CANES
VENATICI
COMA BERENICES
CAM
Arcturus
Polaris
BOÖTES
URSA MINOR
CORONA
BOREALIS
DRACO
SERPENS CAPUT
CEPHEUS
HERCULES
OPH
CYGNUS
ORTH

EAST

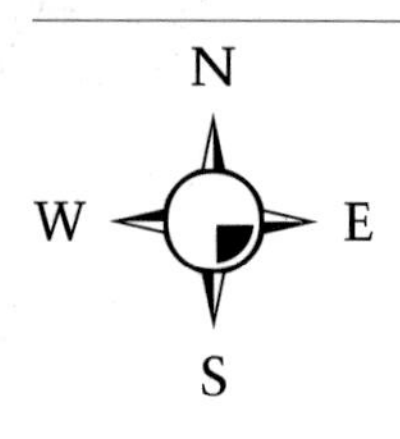

THE SPRING SKY: Southeast

Chart shows: March 15 sky, 11 P.M.
April 15 sky, 9 P.M.; 10 P.M. DST
May 15 sky, 7 P.M.; 8 P.M. DST

The brightest stars of the southeastern sky are bright white Spica, the α star in Virgo; and the red giant Arcturus, the α star in Boötes. They can be found by continuing the line from the handle of the Big Dipper, first to Arcturus and on to Spica. To the right of Spica are the four noticeable stars of Corvus, the Crow. The bright star above Virgo is Denebola, the β star in the tail of Leo, the Lion. To the east of Denebola is the cluster of stars in Coma Berenices, Berenice's Hair, that used to be considered the tuft in Leo's tail. When you look toward Coma Berenices you are looking north, perpendicular to the plane of the Milky Way, and hence have a clear view of the regions beyond our own galaxy. Coma and nearby Virgo house thousands of galaxies, most too faint to be seen with amateur instruments.

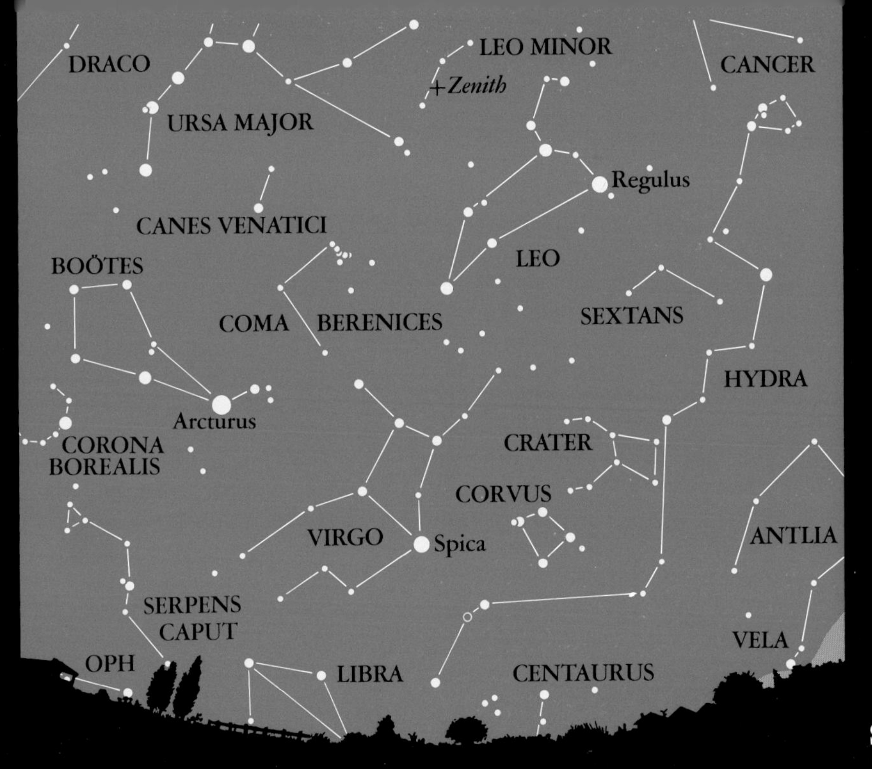
DRACO
LEO MINOR
CANCER
+Zenith
URSA MAJOR
Regulus
CANES VENATICI
BOÖTES
LEO
COMA
BERENICES
SEXTANS
HYDRA
Arcturus
CORONA
BOREALIS
CRATER
CORVUS
VIRGO
Spica
ANTLIA
SERPENS
CAPUT
VELA
OPH
LIBRA
CENTAURUS

EAST

SOUTH

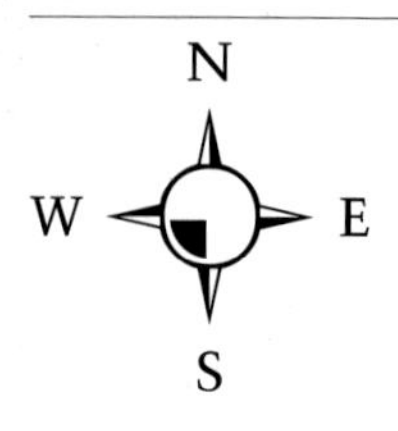

THE SPRING SKY: Southwest

Chart shows: March 15 sky, 11 P.M.
April 15 sky, 9 P.M.; 10 P.M. DST
May 15 sky, 7 P.M.; 8 P.M. DST

The giant figure of Orion is very low in the southwest. Sirius, the Dog Star (α star in Canis Major), is just above the horizon; above Sirius is Procyon (α CMi), the Little Dog Star. Higher still in the west-southwest are the Twins of Gemini, Castor (α Gem) and Pollux (β Gem). Regulus, the α star in Leo, is nearly on the southern meridian. Faint constellations, including Hydra, the Sea Serpent, and Cancer, the Crab, fill the rest of the southwestern sky. Hydra is the largest and longest constellation in the sky and takes more than six hours to rise completely. The serpent's head lies beneath Cancer, a little to the east of Procyon.

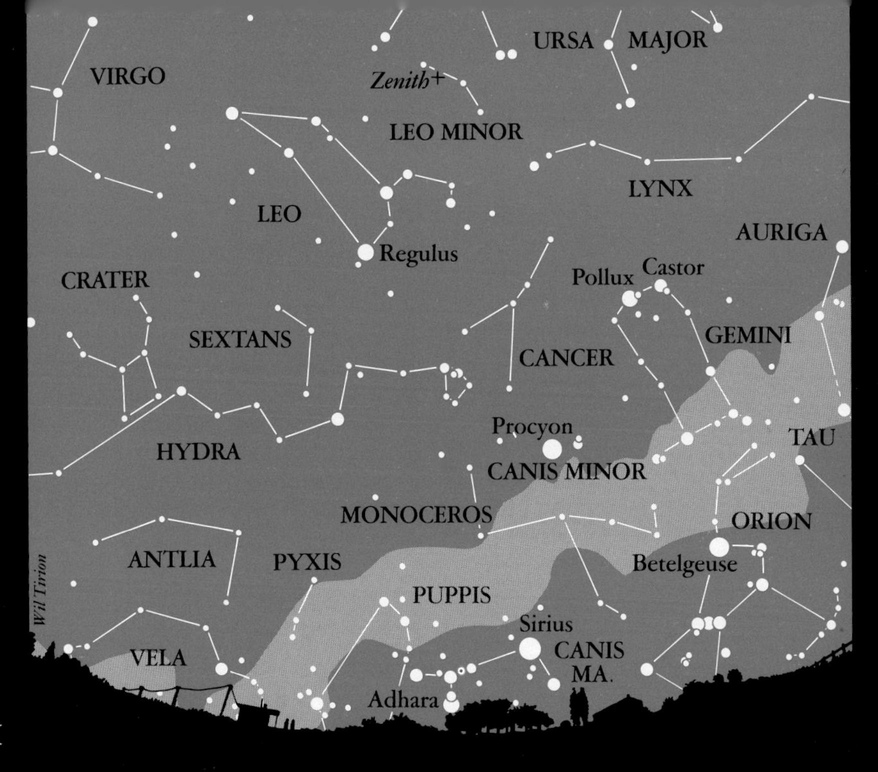

URSA MAJOR
VIRGO
Zenith+
LEO MINOR
LYNX
LEO
Regulus
AURIGA
CRATER
Pollux
Castor
SEXTANS
GEMINI
CANCER
Procyon
TAU
HYDRA
CANIS MINOR
MONOCEROS
ORION
ANTLIA
PYXIS
Betelgeuse
Wil Tirion
PUPPIS
Sirius
CANIS MA.
VELA
Adhara

OUTH

WEST

THE SUMMER SKY: Northwest

Chart shows: June 15 sky, 11 P.M.; midnight DST
July 15 sky, 9 P.M.; 10 P.M. DST
August 15 sky, 7 P.M.; 8 P.M. DST

Ursa Major, with its well-known asterism the Big Dipper, dominates the northwestern sky at this time. The arc of its handle points to Arcturus (the α star in Boötes), the brightest red giant in the sky. Winding between Ursa Major and Ursa Minor (the Little Dipper) are the dim stars of Draco, the Dragon; this is a good time to look for the Dragon as he is highest now, stretching from the northwest above Polaris (α UMi) into the northeast. Leo plunges downward due west; Regulus (α Leo) has already disappeared below the horizon. Above Leo are the faint stars of Coma Berenices, considered by some civilizations to be the tuft at the end of the lion's tail.

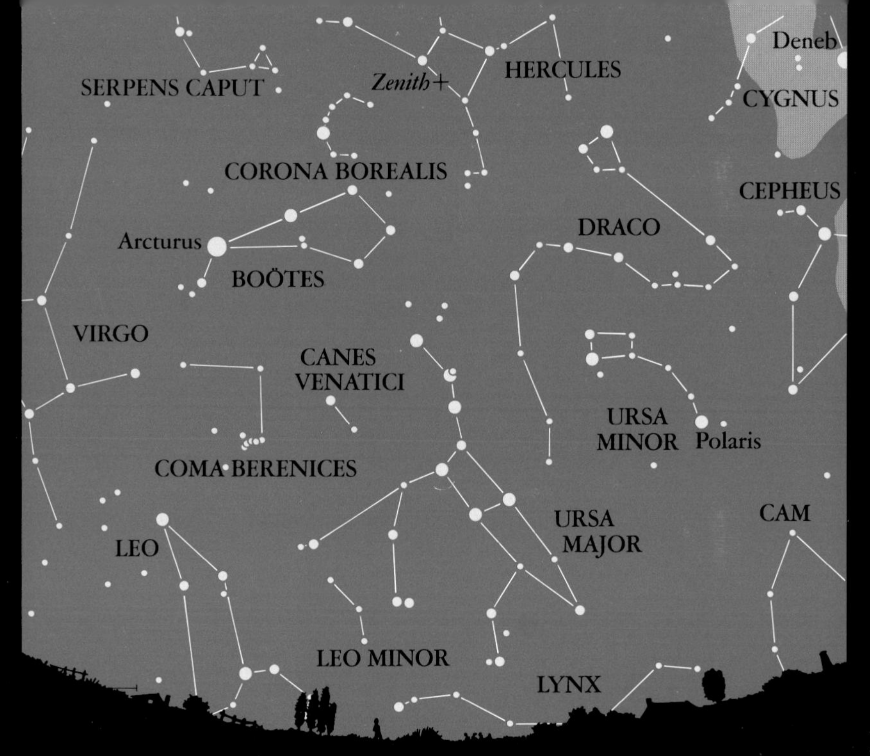
SERPENS CAPUT
Zenith+
HERCULES
Deneb
CYGNUS
CORONA BOREALIS
CEPHEUS
Arcturus
DRACO
BOÖTES
VIRGO
CANES
VENATICI
URSA
MINOR
Polaris
COMA BERENICES
URSA
MAJOR
CAM
LEO
LEO MINOR
LYNX

VEST NORTH

THE SUMMER SKY: Northeast

Chart shows: June 15 sky, 11 P.M.; midnight DST
July 15 sky, 9 P.M.; 10 P.M. DST
August 15 sky, 7 P.M.; 8 P.M. DST

The high northeastern sky is dominated by the three bright stars of the Summer Triangle: Vega, the α star in Lyra, the Harp, is highest; Deneb, the α star in Cygnus, the Swan, is toward the northeast; and Altair, the α star in Aquila, the Eagle, is farther east. The latter two are immersed in the hazy light of the Milky Way, running from the north-northeastern horizon through Cassiopeia and upward. Below Altair is the distinctive, albeit tiny and dim, constellation Delphinus, the Dolphin. To the left of Cygnus is Cepheus, and below him is Cassiopeia. The first stars of the Great Square of Pegasus are just rising above the northeastern horizon, an early sign of fall.

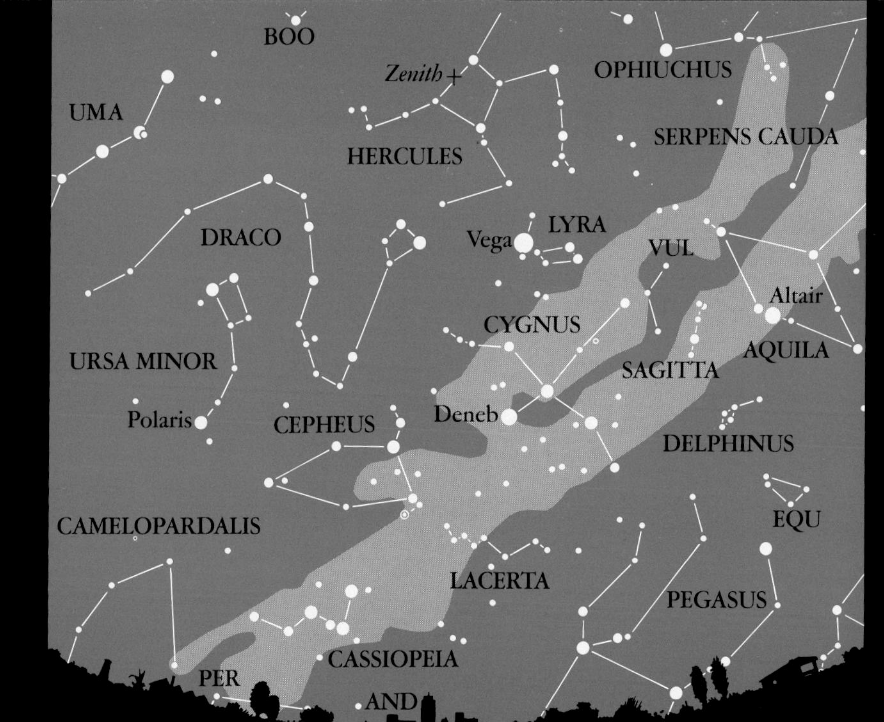

BOO
Zenith
OPHIUCHUS
UMA
SERPENS CAUDA
HERCULES
LYRA
DRACO
Vega
VUL
CYGNUS
Altair
URSA MINOR
AQUILA
SAGITTA
Deneb
Polaris
CEPHEUS
DELPHINUS
CAMELOPARDALIS
EQU
LACERTA
PEGASUS
CASSIOPEIA
PER
AND

THE SUMMER SKY: Southeast

Chart shows: June 15 sky, 11 P.M.; midnight DST
July 15 sky, 9 P.M.; 10 P.M. DST
August 15 sky, 7 P.M.; 8 P.M. DST

Sagittarius, with its asterism known as the Teapot, is low in the southeast. When you look in the direction of Sagittarius you are looking toward the center of our galaxy. Here the star clouds of the Milky Way are brightest, stretching up through the southeastern sky across to the northeast, through Aquila, Cygnus, and Cassiopeia. Altair, the α star in Aquila, is in the east, with bright Vega, the α star in Lyra, above it. Higher still, at the zenith, is H-shaped Hercules. If the sky is really dark, now is a good time to try to find the huge figure of Ophiuchus and the two parts of Serpens that entwine it. These faint stars occupy a large part of the southern sky below Hercules.

DRACO
CORONA BOREALIS
+Zenith
VIRGO
Vega
LYRA
SERPENS
CAPUT
Deneb
HERCULES
CYGNUS
OPHIUCHUS
VULPECULA
SERPENS
CAUDA
LIBRA
SAGITTA
Altair
Antares
DELPHINUS
SCUTUM
AQUILA
LUP
PEGASUS
EQUULEUS
SCORPIUS
NOR
AQUARIUS
SAGITTARIUS
CAPRICORNUS
CRA
TEL
EAST
SOUTH

THE SUMMER SKY: Southwest

Chart shows: June 15 sky, 11 P.M.; midnight DST
July 15 sky, 9 P.M.; 10 P.M. DST
August 15 sky, 7 P.M.; 8 P.M. DST

Kite-shaped Boötes and its α star Arcturus are high in the west. Between Boötes and Hercules, at the zenith, is Corona Borealis, the Northern Crown. Scorpius and its bright red α star Antares are due south. Spica, Virgo's α star, sits alone, low in the west. Between Scorpius and Virgo lie the stars of Libra, the Scales, which were once considered the claws of the Scorpion. Peeking over the southern horizon is one of the gems of the southern sky, Centaurus. Its brightest stars, Rigel Kentaurus (α) and Agena (β), however, are too low to be seen from most northern locations. Both are among the brightest stars in the sky, and α Centauri, a triple star, is the nearest star system to Earth, only 4.3 ly away.

+Zenith
SERPENS
CAUDA
HERCULES
URSA MAJOR
BOÖTES
CORONA
BOR.
CANES VENATICI
SCT
OPHIUCHUS
Wil Tirion
SERPENS
CAPUT
Arcturus
COMA
BERENICES
Antares
VIRGO
SCORPIUS
LIBRA
LEO
Spica
LUPUS
HYDRA
NORMA
CORVUS
CENTAURUS
CRATER

THE AUTUMN SKY: Northwest

Chart shows: September 15 sky, 11 P.M.; midnight DST
October 15 sky, 9 P.M.; 10 P.M. DST
November 15 sky, 7 P.M.

The Summer Triangle can still be seen in the northwestern sky: Altair, the α star in Aquila, the Eagle, is almost due west; Vega, the α star in Lyra, the Harp, is about halfway up the northwestern sky; and Deneb, the α star in Cygnus, the Swan, is about two-thirds of the way between the horizon and the zenith. Low in the northwest is Hercules. Ursa Major and its well-known asterism the Big Dipper are very low on the northern horizon. Ursa Minor, the Little Dipper, now extends to the left of Polaris (α UMi), the North Star, which marks the end of its handle. Above the Little Dipper is Cepheus, the King to Cassiopeia's Queen and father of Princess Andromeda, both of whom are to the east. The stars of Draco, the Dragon, run between the Little Dipper and Hercules.

PEGASUS
ANDROMEDA
EQUULEUS
+ Zenith
PERSEUS
LACERTA
DELPHINUS
AQUILA
Altair
CASSIOPEIA
SAGITTA
Deneb
CAM
CYGNUS
CEPHEUS
VUL
Polaris
SERPENS
CAUDA
LYRA
Vega
URSA MINOR
OPHIUCHUS
HERCULES
DRACO
CORONA
BOR.
BOÖTES
URSA MAJOR
WEST
NORTH

N

W E

S

THE AUTUMN SKY: Northeast

Chart shows: September 15 sky, 11 P.M.; midnight DST
October 15 sky, 9 P.M.; 10 P.M. DST
November 15 sky, 7 P.M.

In the northeast is the W shape of Cassiopeia, the Queen. Her daughter Andromeda, beside her, stretches from the northeastern star of the Great Square of Pegasus toward the horizon. Below Cassiopeia and Andromeda is Perseus, the Hero. The brightest stars in this part of the sky are below this royal group: the yellowish star Capella (α Aur), in the pentagonal constellation Auriga, the Charioteer, and the reddish star Aldebaran (α Tau), the eye of Taurus, the Bull, low in the east. The V-shaped star cluster of the Hyades (with Aldebaran at the end of the bottom arm of the V) at this time looks like an arrowhead pointing to the right. Above Aldebaran lies the small but beautiful star cluster of the Pleiades, an interesting sight in binoculars. The Milky Way runs perpendicular to the horizon here, from Auriga up through Cygnus and beyond.

Deneb
CYGNUS
Zenith
PEGASUS
PISCES
LACERTA
ANDROMEDA
DRACO
CASSIOPEIA
CEPHEUS
TRIANGULUM
ARIES
Polaris
URSA MINOR
Algol
CETUS
CAMELOPARDALIS
Pleiades
PERSEUS
TAURUS
Capella
URSA MAJOR
Aldebaran
AURIGA
ORI
LYNX
ORTH
EAST

THE AUTUMN SKY: Southeast

Chart shows: September 15 sky, 11 P.M.; midnight DST
October 15 sky, 9 P.M.; 10 P.M. DST
November 15 sky, 7 P.M.

This is not a particularly bright part of the sky. Thousands of years ago it was known as "the Water" or "the Sea" and is filled with aquatic constellations, such as Pisces, the Fish; Cetus, the Sea Monster; Aquarius, the Waterbearer; Piscis Austrinus, the Southern Fish; and Eridanus, the River. Although Cetus and Pisces occupy a huge area of the sky, they are fairly uninteresting for the casual stargazer. Higher in the southeast are the stars of Pegasus, the Winged Horse, and its asterism the Great Square. The curved A-shape of Andromeda, the Princess, begins at the northeastern corner of the Great Square and stretches into the northeastern sky.

CASSIOPEIA
LACERTA
DELPHINUS
Zenith
ANDROMEDA
PERSEUS
PEGASUS
EQUULEUS
Algol
AQUARIUS
TRIANGULUM
PISCES
ARIES
CAPRICORNUS
Pleiades
Fomalhaut
Mira
PISCIS
AUST.
TAURUS
SCULPTOR
GRUS
CETUS
ERIDANUS
PHOENIX
FORNAX
EAST
SOUTH

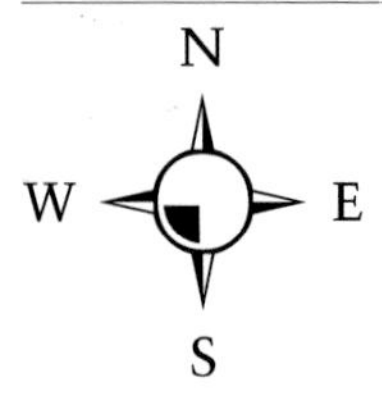

THE AUTUMN SKY: Southwest

Chart shows: September 15 sky, 11 P.M.; midnight DST
October 15 sky, 9 P.M.; 10 P.M. DST
November 15 sky, 7 P.M.

The Water or Sea extends into the southwestern portion of the autumn sky, with Capricornus, the Sea Goat, and Delphinus, the Dolphin, joining Aquarius. Just about due south is one of the brighter stars in the sky, Fomalhaut, the α star in Piscis Austrinus, the Southern Fish. Fomalhaut is unfamiliar to most northern-latitude observers as it appears in northern skies only briefly and is always low. There are no other bright stars in the southern sky at this time, however, so it is unmistakable. Sagittarius is setting in the southwestern sky; it will not be seen again until late next spring. The Milky Way runs up from the horizon through Sagittarius, Aquila, and Cygnus.

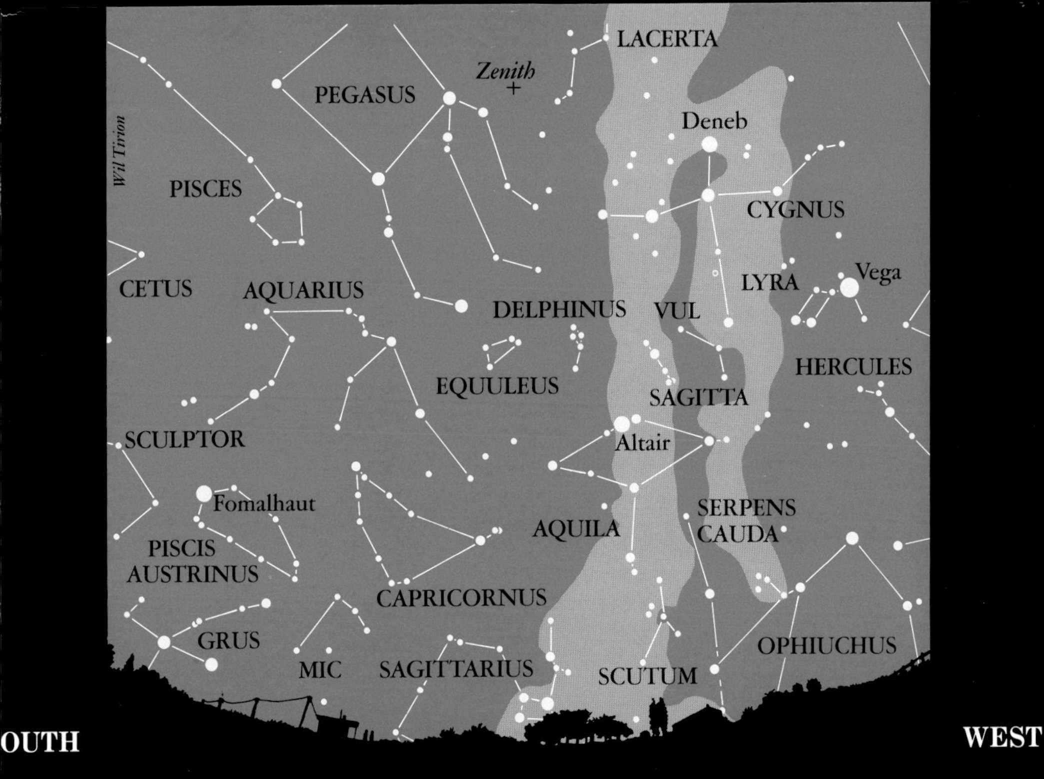

LACERTA
Zenith
PEGASUS
Deneb
Wil Tirion
PISCES
CYGNUS
LYRA
Vega
CETUS
AQUARIUS
DELPHINUS
VUL
EQUULEUS
HERCULES
SAGITTA
SCULPTOR
Altair
Fomalhaut
SERPENS
CAUDA
AQUILA
PISCIS
AUSTRINUS
CAPRICORNUS
GRUS
OPHIUCHUS
MIC
SAGITTARIUS
SCUTUM
WEST

SERPENS
SERPEN
Æquat

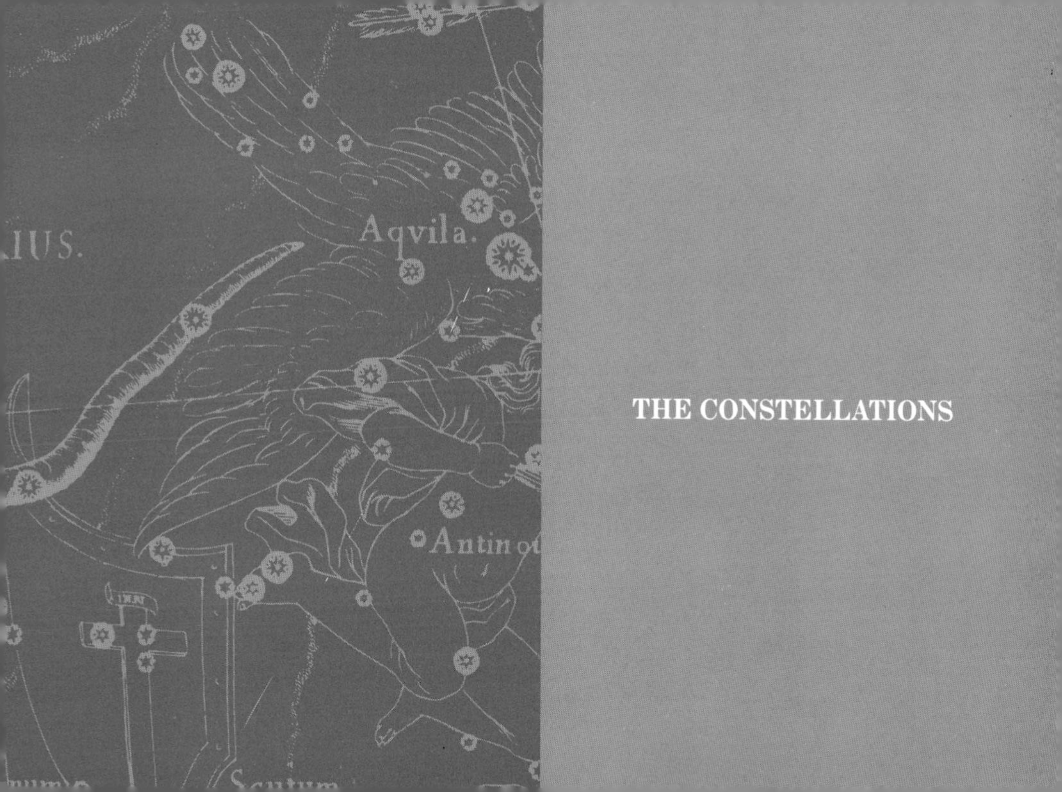

THE CONSTELLATIONS

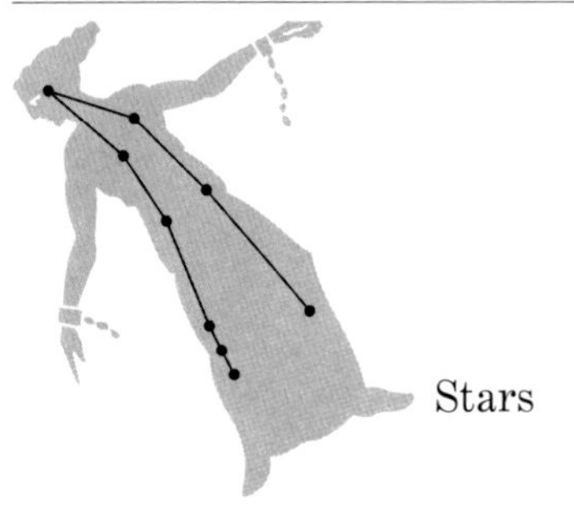

Andromeda (And) The Princess, the Chained Maiden
On Meridian: November 10   Area: 722 square degrees

Princess Andromeda, the daughter of Cepheus and Cassiopeia, was chained to a rock and left to be devoured by Cetus, the sea monster. She was rescued by Perseus, slayer of the Gorgons.

Stars

α Andromedae, called Alpheratz or Sirrah, marks one corner of the Great Square of Pegasus (the three brightest stars of Pegasus mark the others). Spectral type B9 IV; magnitude 2.1; distance 100 ly.

Deep-Sky Objects

The famous Andromeda Galaxy (M31, NGC 224), at 2.5 million ly, is the most distant object visible to the naked eye. It is a spiral galaxy, the largest member of the Local Group, the small cluster of galaxies to which our Milky Way belongs. It appears to the unaided eye on clear, dark nights as a faint oval of fuzzy light. Its two bright companion galaxies are M32 (NGC 221), a round, 9th-magnitude dwarf galaxy next to M31, and M110 (NGC 205), a more flattened, dwarf elliptical galaxy slightly farther out. A telescope is needed to see them. NGC 752 is an open cluster visible in binoculars.

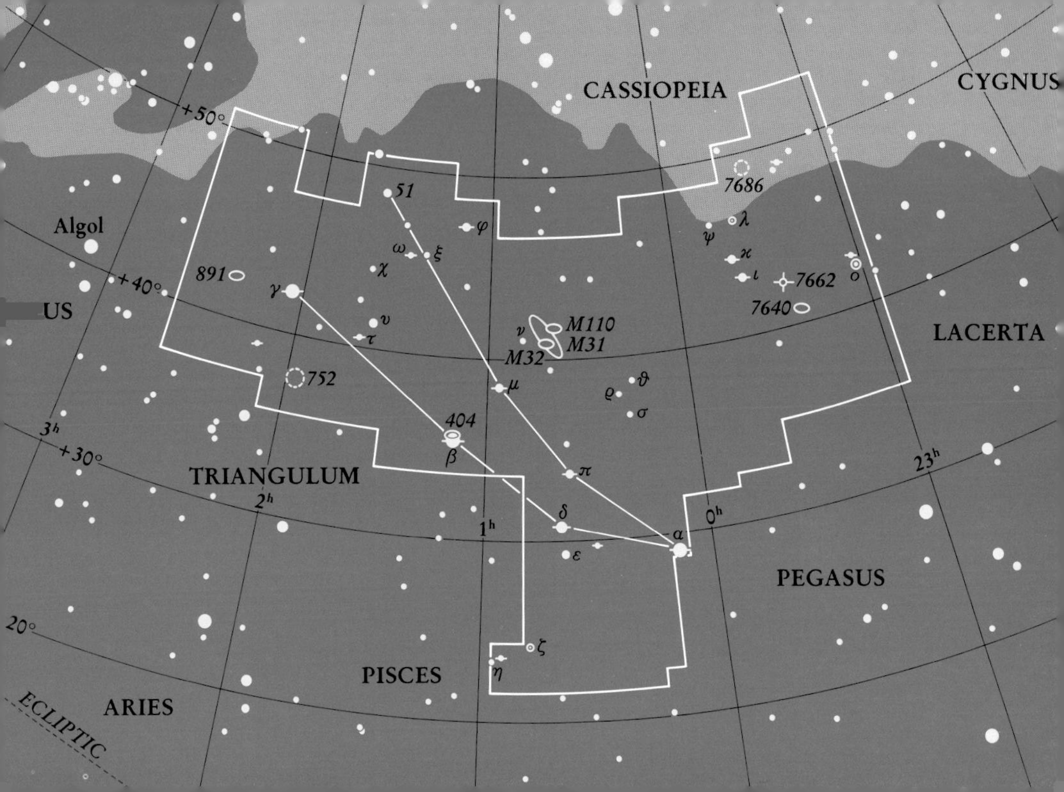
CASSIOPEIA
CYGNUS
+50°
+40°
+30°
20°
Algol
US
LACERTA
TRIANGULUM
PEGASUS
PISCES
ARIES
ECLIPTIC
3h
2h
1h
0h
23h
51
φ
ω
ξ
χ
γ
891
υ
τ
752
404
β
ν
M110
M31
M32
μ
ϑ
ϱ
σ
π
δ
α
ε
ζ
η
7686
λ
ψ
ϰ
ι
7662
o
7640

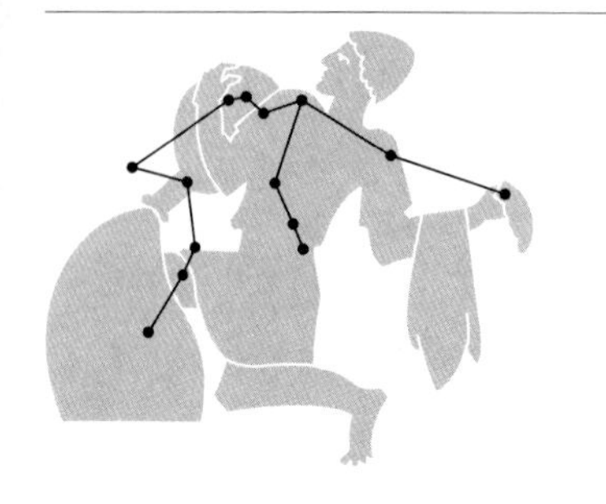

Aquarius (Aqr) The Waterbearer
On Meridian: October 10   Area: 980 square degrees

One of the zodiacal constellations, Aquarius is also one of the most ancient constellations in the sky and has been known under various names over the ages. The constellation ostensibly portrays a man or boy spilling water from an urn. It figured in much ancient mythology, including a Sumerian myth of a global deluge that is thought to be the antecedent to the biblical flood story.

Stars

α, named Sadalmelik (Arabic for "lucky one of the king"), lies almost exactly on the celestial equator. At only 3.0 magnitude, it is the brightest star of this dim constellation. Spectral type G2 Ib; distance 680 ly.

Deep-Sky Objects

There is a notable globular cluster, M2, of about 7th magnitude, located a few degrees north of β Aquarii. Two planetary nebulae are nearby. The larger is NGC 7293, the well-known Helix Nebula, located southwest of Skat (δ Aqr). NGC 7009, of 8th magnitude, called the Saturn Nebula for its appearance, is southeast of Albali (ϵ Aqr).

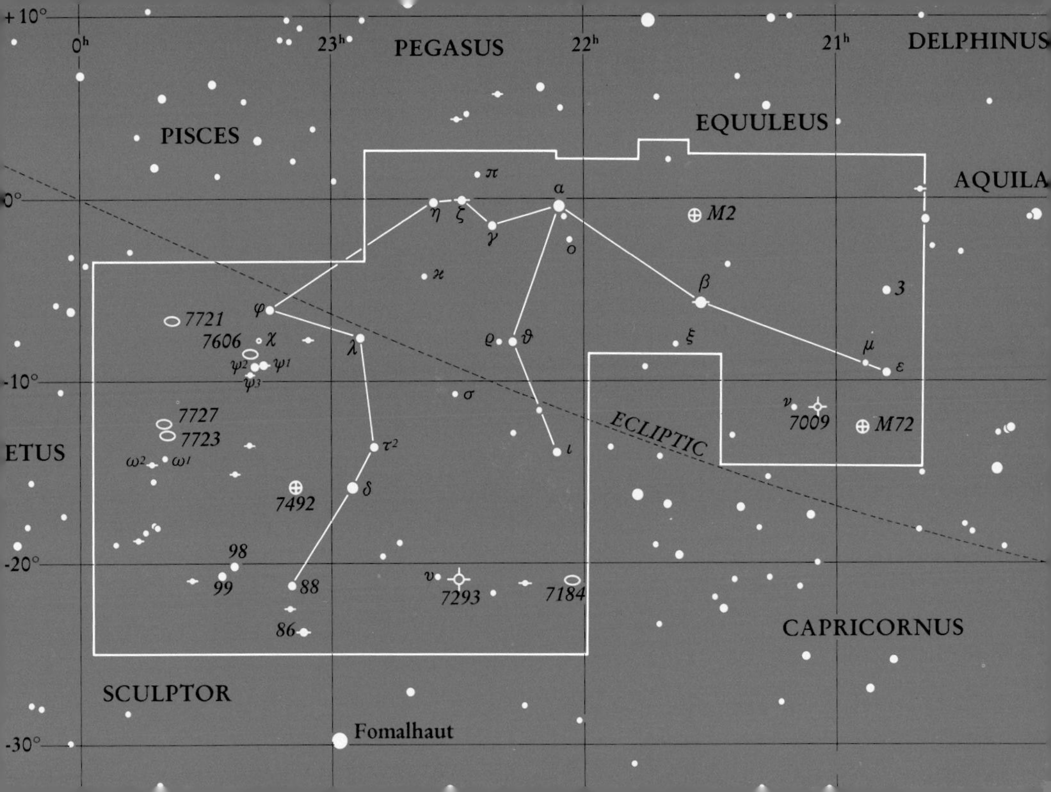
+10°
0°
-10°
-20°
-30°
0h
23h
22h
21h
PEGASUS
DELPHINUS
PISCES
EQUULEUS
AQUILA
ETUS
SCULPTOR
CAPRICORNUS
ECLIPTIC
Fomalhaut
M2
M72
7009
7721
7606
7727
7723
7492
7293
7184
98
99
88
86
3

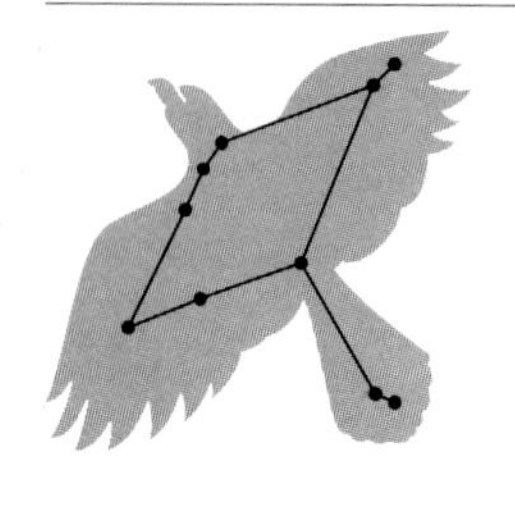

Aquila (Aql) The Eagle
On Meridian: August 30   Area: 652 square degrees

Aquila, identified as a bird since around 1200 B.C. in Western tradition, lies along the summer Milky Way and is a rich area to explore with a wide-field telescope.

Stars

α, Altair, is among the brightest stars in the sky, and is the middle star in the line of three that forms the wings of the eagle. It marks one corner of the Summer Triangle, a bright shape visible in the northern sky in summer, the other points of which are formed by Deneb (α Cyg) and Vega (α Lyr). Spectral type A7 IV–V; magnitude 0.8; distance 17 ly.

Deep-Sky Objects

NGC 6709 is an 8th-magnitude open cluster of about 40 stars, located a few degrees southwest of ζ Aquilae, at a distance of 3,100 ly from Earth. A small telescope is needed to see it.

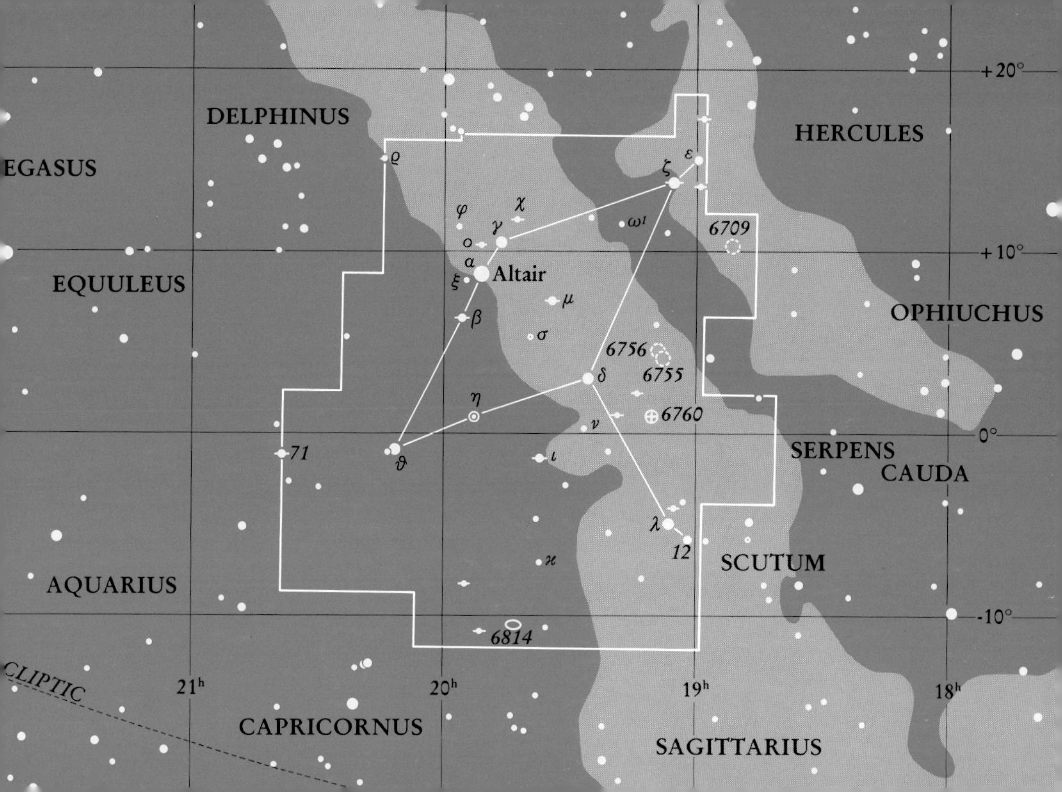

DELPHINUS
EGASUS
EQUULEUS
HERCULES
OPHIUCHUS
Altair
6709
6756
6755
6760
71
SERPENS
CAUDA
SCUTUM
AQUARIUS
12
6814
CLIPTIC
21h
20h
19h
18h
+20°
+10°
0°
-10°
CAPRICORNUS
SAGITTARIUS

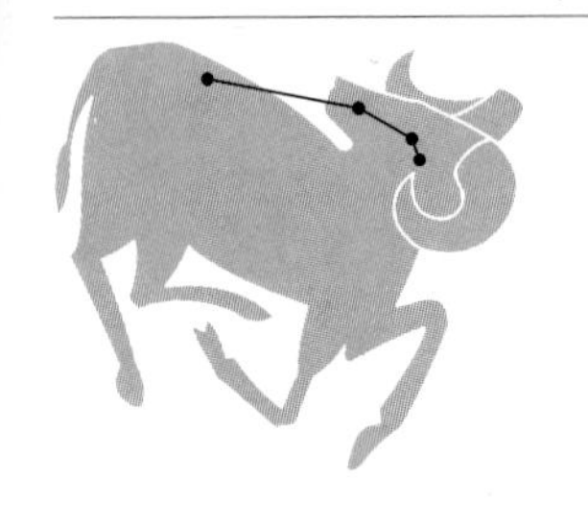

Aries (Ari) The Ram

On Meridian: December 10   Area: 441 square degrees

The second-smallest constellation of the zodiac, Aries is composed principally of only three lackluster stars that make a very flat scalene triangle. Aries claims a history and mythology that go back far in time. For much of recorded history, the Sun in Aries marked the beginning of spring. Because of precession, however, the Sun is now in neighboring Pisces at the vernal equinox.

Stars

α is a red giant called Hamal ("lamb"), the Arabic name for the entire constellation. Spectral type K2 III; magnitude 2.0; distance 78 ly. β is called Sheratan, Arabic for "two of something," although two of what is not clear. Spectral type A4 V; magnitude 2.6; distance 44 ly. γ, named Mesarthim, appears to the unaided eye as a single star of about 4th magnitude, but a small telescope reveals a pair of stars of magnitude 4.6, both type A0 V, about 115 ly away.

Deep-Sky Objects

None of easy viewing or noteworthiness.

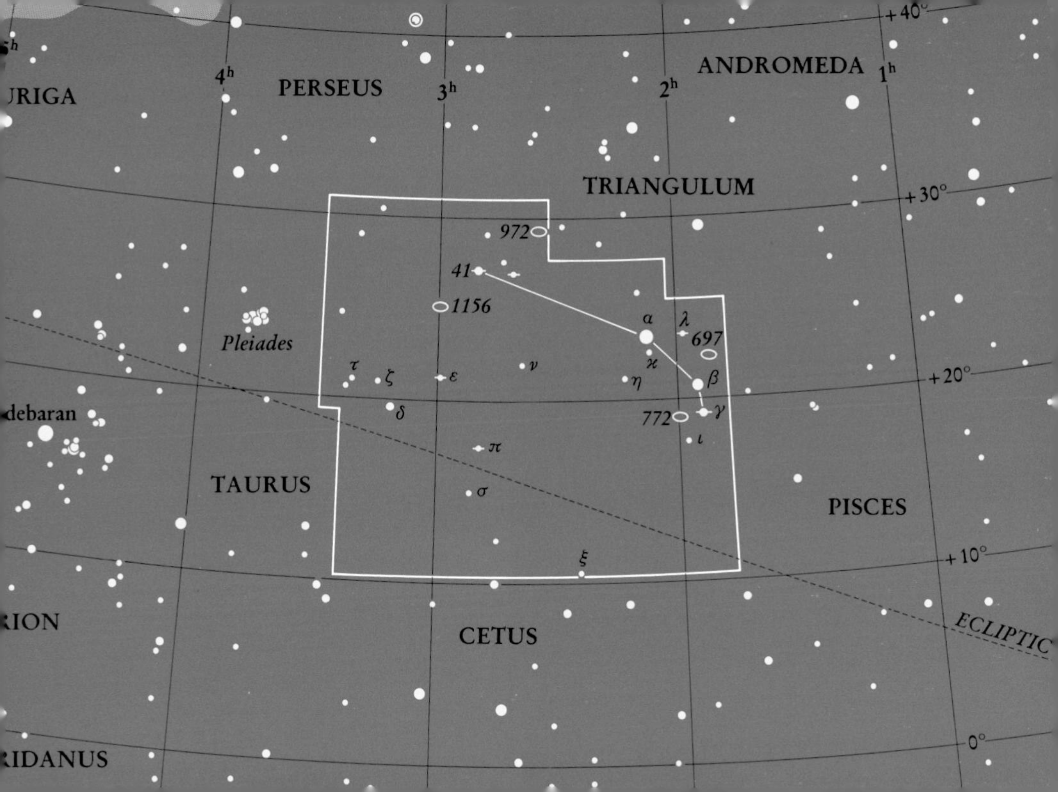

+40°
ANDROMEDA
1h
PERSEUS
4h
3h
2h
URIGA
TRIANGULUM
+30°
972
41
1156
Pleiades
α
λ
697
τ
ζ
ε
ν
κ
η
β
+20°
debaran
δ
772
γ
ι
π
TAURUS
σ
PISCES
ξ
+10°
RION
CETUS
ECLIPTIC
0°
RIDANUS

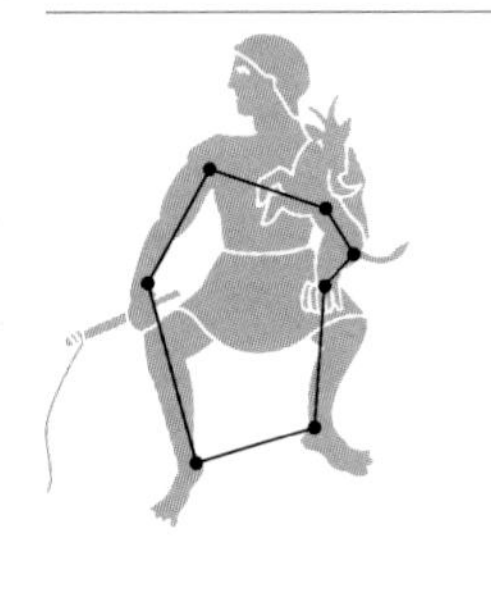

Auriga (Aur) The Charioteer
On Meridian: January 10   Area: 657 square degrees

Auriga, one of the oldest constellations, is associated in Greek mythology with the lame inventor of the chariot. He is usually portrayed as a goatherd, holding a kid.

Stars

α, a yellowish giant star named Capella, is the sixth-brightest star in the sky. When rising in the early evening in the northeastern sky, Capella signals the beginning of autumn and the imminent rising of the winter constellations Taurus and Orion. Spectral type G6 III; magnitude 0.1; distance 42 ly.

Deep-Sky Objects

Situated along the Milky Way, Auriga offers four easily visible star clusters. M36, a large open cluster of about 6th magnitude, lies 3,700 ly away and contains about 60 stars. M37, another large, 6th-magnitude open cluster, contains about 150 stars. M38 is a 7th-magnitude cluster of about 100 stars. NGC 2281, also of about 7th magnitude, contains about 30 stars in an area approximately ⅙° across.

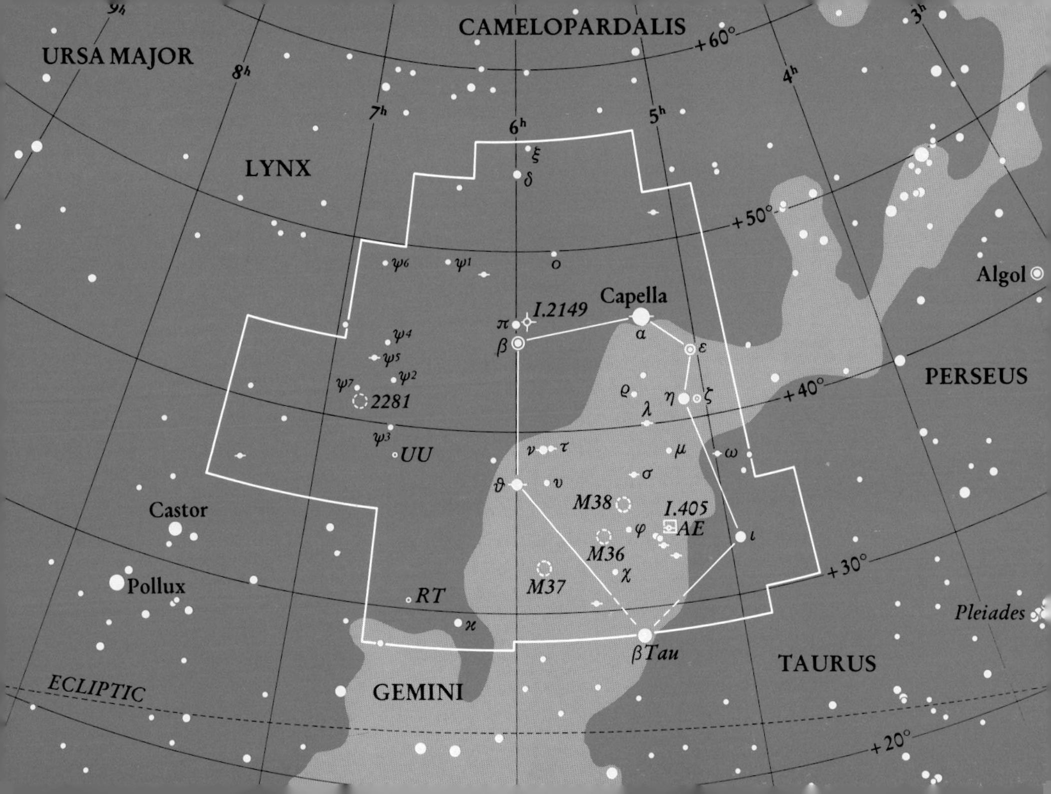
URSA MAJOR
CAMELOPARDALIS
LYNX
PERSEUS
TAURUS
GEMINI
ECLIPTIC
Capella
Castor
Pollux
Algol
Pleiades
βTau
I.2149
I.405
AE
2281
UU
RT
M38
M36
M37
+60°
+50°
+40°
+30°
+20°
9h
8h
7h
6h
5h
4h
3h

Boötes (Boo) The Herdsman
On Meridian: June 15 Area: 907 square degrees

The constellation of Boötes is sometimes depicted as a bear-keeper leading the dogs of Canes Venatici and chasing the bears of Ursa Major and Ursa Minor around the north celestial pole. Boötes has the shape of a kite or an ice cream cone, with its brightest star, Arcturus, at the bottom. To find Arcturus, follow the curving handle of the Big Dipper and then "arc to Arcturus."

Stars

α is Arcturus, from the Greek *arktouros,* meaning "guardian of the bear," referring to the great bear of Ursa Major. It is the fourth-brightest star in the entire sky, and one of the few stars bright enough for its color—yellow-orange—to be apparent. Spectral type K2 III; magnitude −0.04; distance 34 ly. The other stars of Boötes are rather faint. The brightest of these, ε, called Izar, is a triple star.

Deep-Sky Objects

NGC 5466, a globular cluster 47,000 ly away, appears as a fuzzy star of magnitude 8.5.

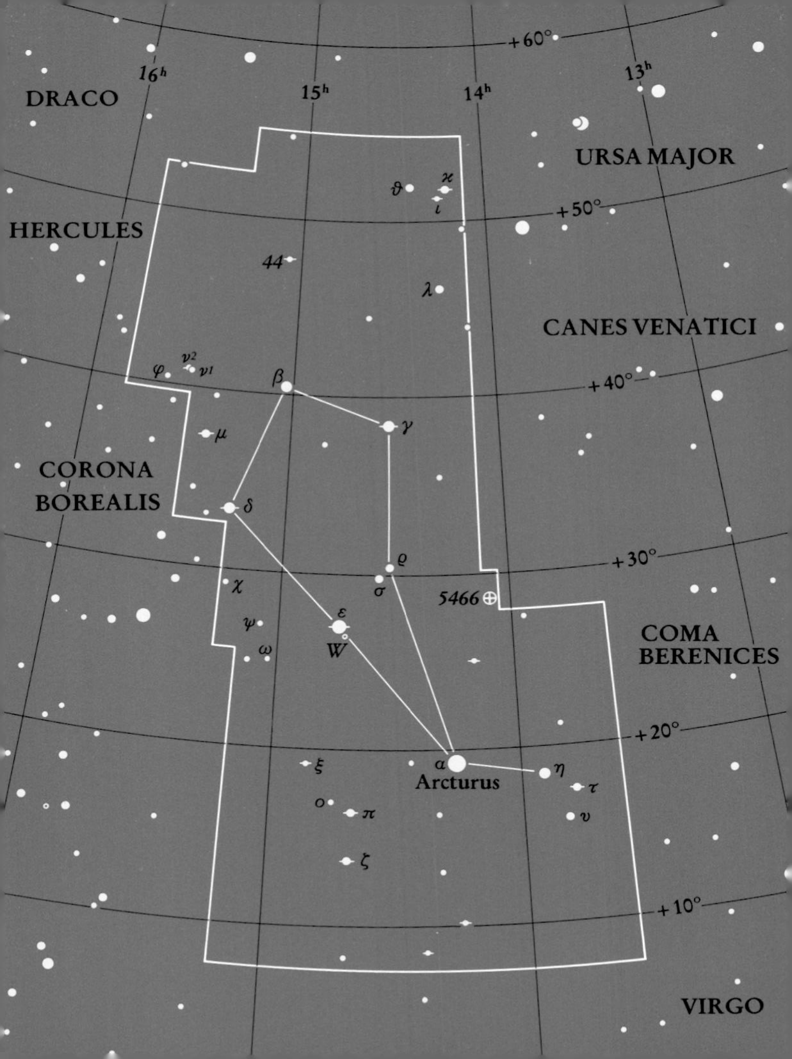
DRACO
16h
15h
14h
13h
+60°
URSA MAJOR
HERCULES
+50°
CANES VENATICI
+40°
CORONA
BOREALIS
+30°
5466
COMA
BERENICES
+20°
Arcturus
+10°
VIRGO

Camelopardalis (Cam) The Giraffe
On Meridian: February 1   Area: 757 square degrees

Camelopardalis, a modern constellation in the far northern sky, was first outlined in 1624 by the German astronomer Jakob Bartsch, who created it to fill a vast region of faint stars surrounded by brighter and more famous constellations (Ursa Major, Cassiopeia, and others). The "camel-leopard" was the name the Greeks gave to the giraffe, an animal that seemed to have the head of a camel and the spots of a leopard. Camelopardalis is among the larger constellations in area, but its brightest stars are faint and unnamed. The Perseid meteor shower of mid-August radiates from its border with Perseus.

Stars — None noteworthy.

Deep-Sky Objects — NGC 1502 is an open cluster, 3,750 ly away, containing about 15 stars, the brightest of which is of about 5th magnitude. NGC 2403 is an 8th-magnitude spiral galaxy with a maximum dimension of about ¼°, or about half the width of the full moon.

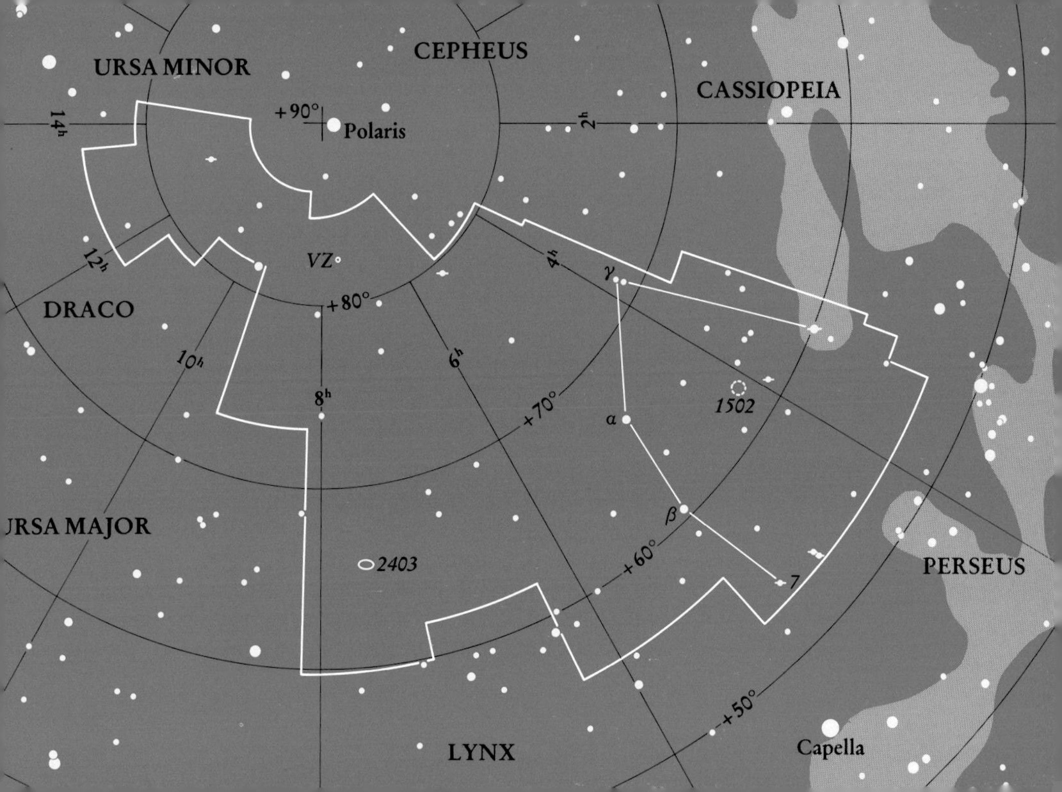

URSA MINOR
CEPHEUS
CASSIOPEIA
+90°
Polaris
14h
2h
12h
4h
VZ
γ
DRACO
+80°
10h
6h
8h
+70°
α
1502
URSA MAJOR
β
+60°
2403
7
PERSEUS
+50°
LYNX
Capella

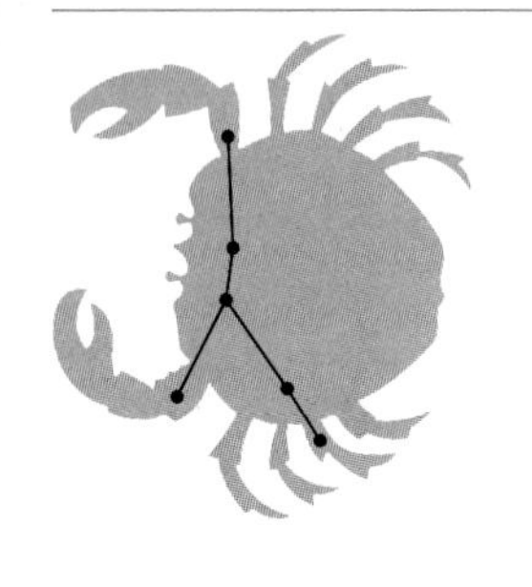

Cancer (Cnc) The Crab

On Meridian: March 15   Area: 506 square degrees

Despite being the least conspicuous constellation of the zodiac, Cancer is one of the best known. In early historical times the Sun was in Cancer at the summer solstice, when it reaches its northernmost position in our sky and is directly overhead at noon at 23.4° N latitude). This line of latitude was thus named the Tropic of Cancer.

Stars

ζ is a well-known multiple star with a combined magnitude of about 4.7, just 52 ly from Earth. In a small telescope two yellow-orange stars, of 5th and 7th magnitudes, are visible.

Deep-Sky Objects

M44 is the famous Praesepe open cluster (also called the Beehive), 515 ly away and visible to the naked eye. It contains about 75 visible stars, the brightest of magnitude 6.3, and extends about 80′ (three times the diameter of the full moon). M67 is another well-studied open cluster of some 65 stars, 15′ across. It has a combined magnitude of 7 and lies about 2,700 ly away.

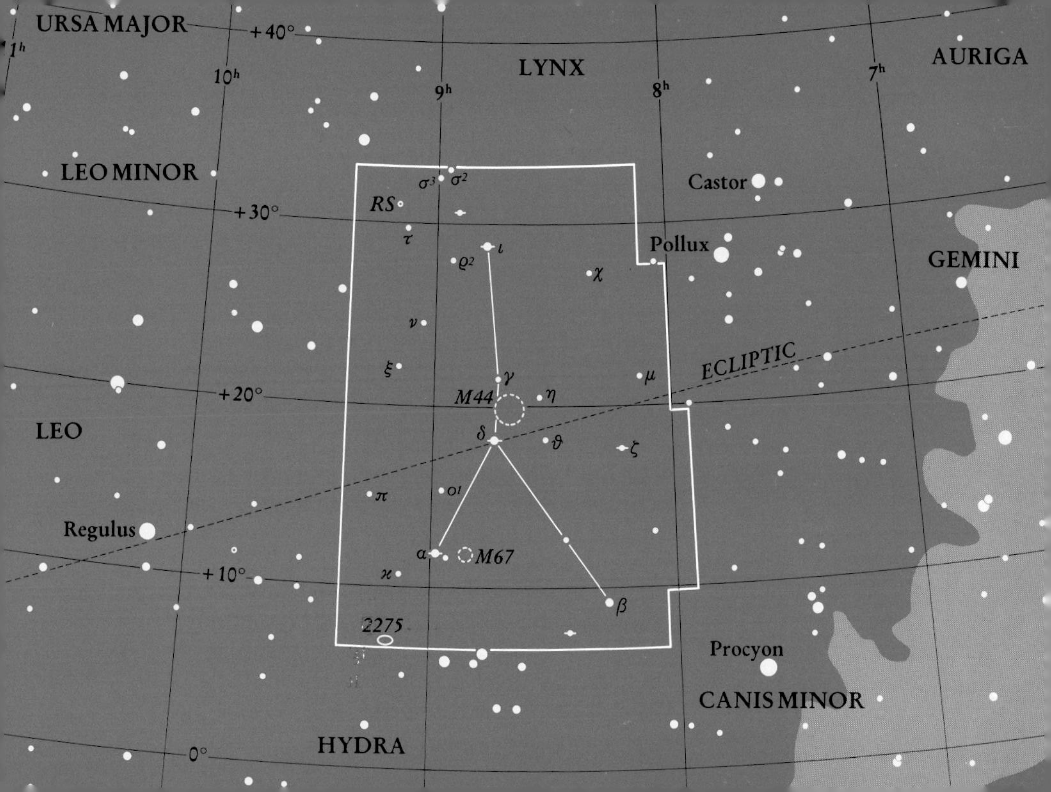

URSA MAJOR
+40°
11h
LYNX
AURIGA
10h
9h
8h
7h
LEO MINOR
Castor
σ3
σ2
RS
+30°
τ
Pollux
ι
ρ2
χ
GEMINI
ν
ECLIPTIC
ξ
γ
μ
M44
η
+20°
LEO
δ
ϑ
ζ
ο1
π
Regulus
α
M67
+10°
κ
β
2275
Procyon
CANIS MINOR
HYDRA
0°

Canes Venatici (CVn) The Hunting Dogs
On Meridian: May 20   Area: 465 square degrees

The hunting dogs of the adjacent herdsman Boötes, Canes Venatici is often portrayed as a pair of greyhounds, leashed and in pursuit of the bears, Ursa Major and Minor. Canes Venatici is a modern constellation, conceived about 1687 by Johannes Hevelius.

Stars

α, the brightest star of Canes Venatici, is known as Cor Caroli, "Charles's heart," a tribute to the British king Charles II. It is a double star, composed of stars of types A0 V and F0 V, of magnitudes 2.9 and 5.5, respectively. The pair is 130 ly from Earth.

Deep-Sky Objects

M3, one of the most splendid globular clusters in the sky, has a diameter of about 10′ and appears as a fuzzy 6th-magnitude star. In a 6″ telescope hundreds of individual stars are revealed. The Whirlpool Galaxy (M51), in the northeastern corner of the constellation (not far from the end of the Big Dipper's handle) appears in a small telescope as an 8th-magnitude fuzzy ball about 10′ in diameter. Its 11th-magnitude companion galaxy, NGC 5195, can also be resolved.

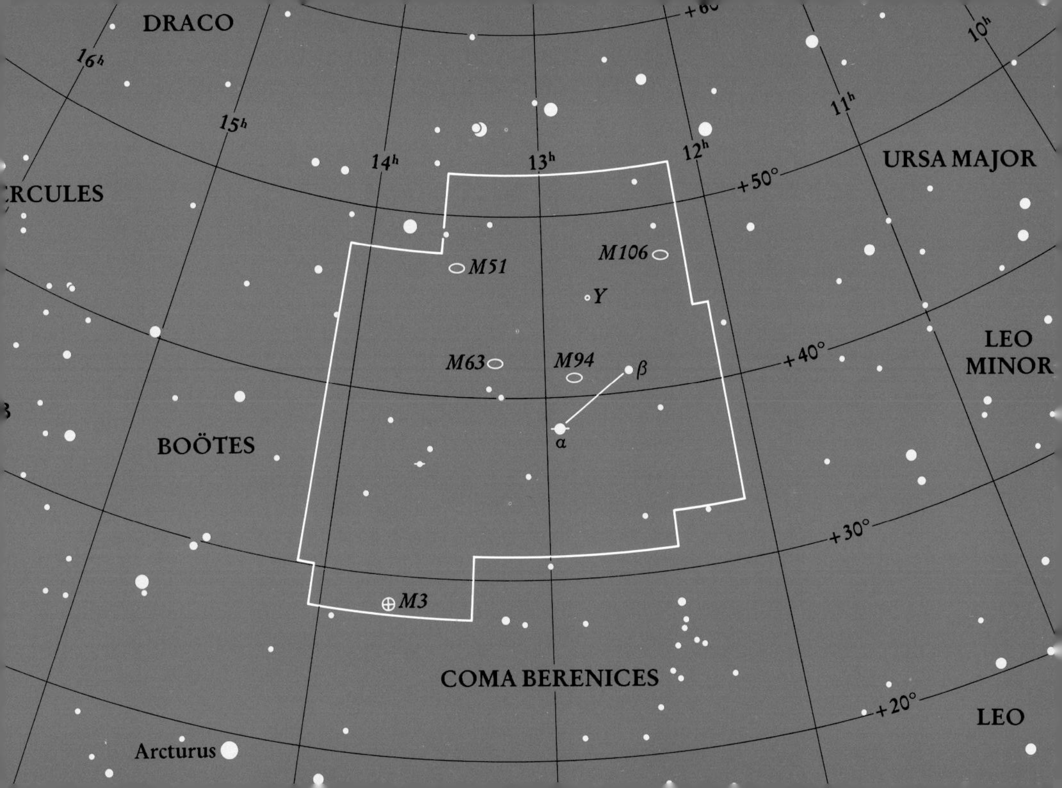

DRACO
16h
15h
14h
13h
12h
11h
10h
+60°
+50°
+40°
+30°
+20°
URSA MAJOR
HERCULES
M106
M51
Y
LEO
MINOR
M63
M94
β
α
BOÖTES
M3
COMA BERENICES
LEO
Arcturus

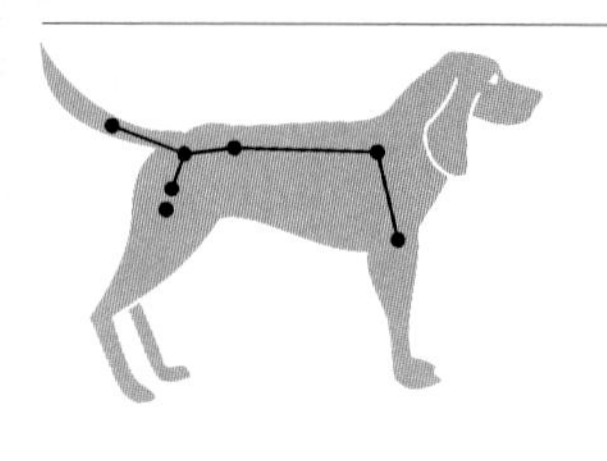

Canis Major (CMa) The Great Dog
On Meridian: February 15 Area: 380 square degrees

Originally the name Canis Major referred only to Sirius, the brightest star in the night sky, said to represent one of Orion's dogs. The Egyptians celebrated the Dog Star's rising with the Sun as the beginning of the year, for it coincided with the annual flooding—and refertilization—of the Nile River.

Stars

α, Sirius, the Dog Star, is a bluish-white star of spectral type A1 V, only 8.8 ly away. This very close proximity (it is the fifth-closest star to our solar system), along with its energy output of about 25 times that of the Sun, makes Sirius the brightest star in the night sky, with an apparent magnitude of −1.46. Sirius has a close, faint, white-dwarf companion (called the Pup) that orbits it once every 50 years. It is at times resolvable in a telescope in good seeing conditions.

Deep-Sky Objects

M41 is a notable open cluster about 4° south of Sirius, visible to the unaided eye as a fuzzy 6th-magnitude star. It contains about 50 7th-magnitude stars within an area about the size of the full moon.

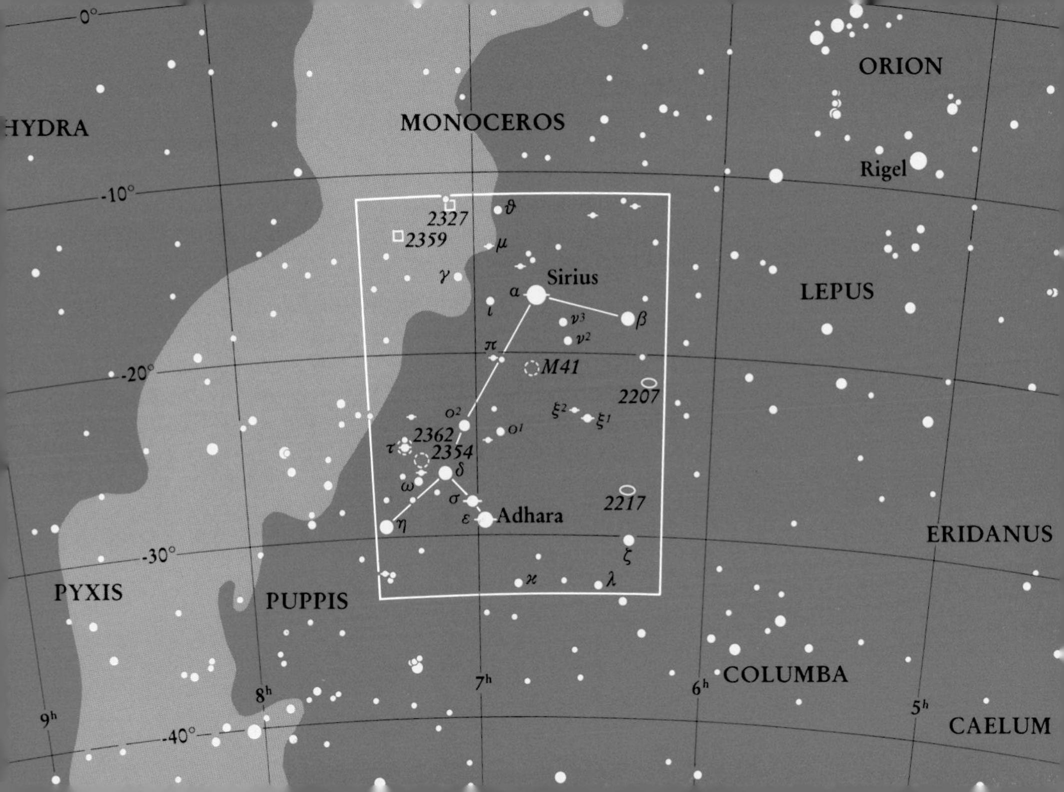

0°
-10°
-20°
-30°
-40°
HYDRA
MONOCEROS
ORION
Rigel
LEPUS
ERIDANUS
PYXIS
PUPPIS
COLUMBA
CAELUM
2327
2359
ϑ
μ
γ
Sirius
α
β
ι
ν3
ν2
π
M41
2207
ξ2
ξ1
o2
o1
2362
τ
2354
δ
ω
σ
ε
Adhara
η
2217
ζ
κ
λ
9h
8h
7h
6h
5h

Canis Minor (CMi) The Little Dog
On Meridian: March 1   Area: 182 square degrees

Canis Minor, the companion of Canis Major, is the other hound of Orion in the winter sky. It is among the smaller constellations in the celestial sphere, basically consisting of just three stars, including the bright Procyon. The Milky Way runs between Sirius and Procyon.

Stars

α is Procyon, from the Greek meaning "before the dog," because it rises before nearby Sirius, the Dog Star. Procyon is only 11.4 ly away, the 14th-nearest star known and the 5th-nearest of the stars visible to the naked eye. Spectral type F5 IV; magnitude 0.4. Procyon, like Sirius, has a white-dwarf companion, too close to the main star and too faint to be seen except in a moderate telescope.

Deep-Sky Objects

None of easy visibility.

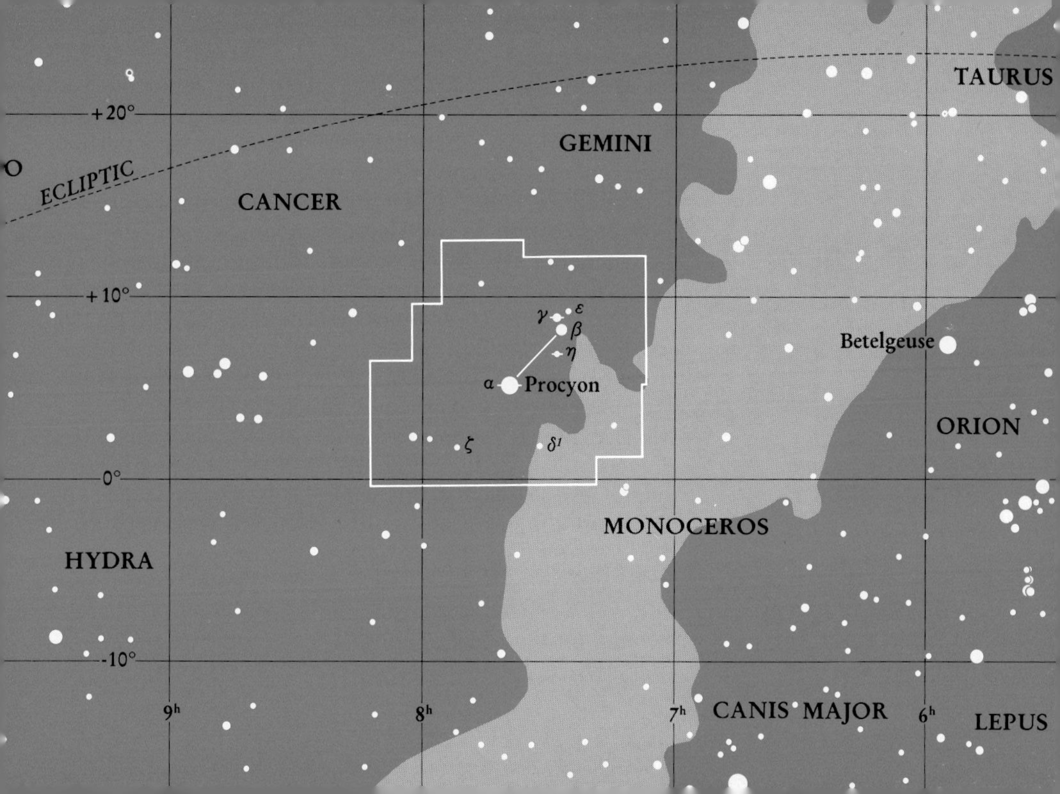

TAURUS
+20°
GEMINI
ECLIPTIC
CANCER
+10°
γ
ε
β
η
Betelgeuse
α
Procyon
ORION
ζ
δ1
0°
MONOCEROS
HYDRA
-10°
9h
8h
7h
CANIS MAJOR
6h
LEPUS

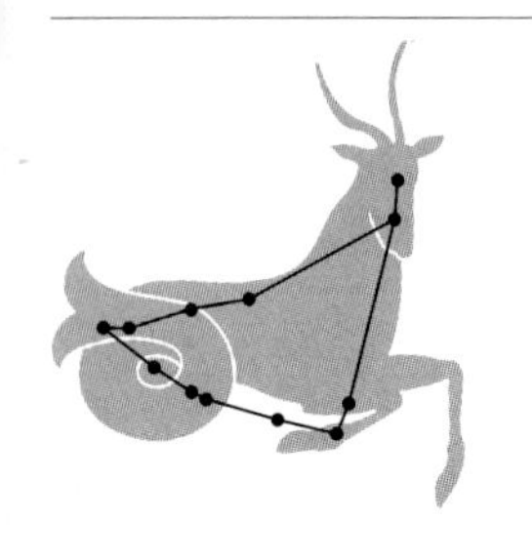

Capricornus (Cap) The Sea Goat, the Goat
On Meridian: September 20 Area: 414 square degrees

The dim zodiacal constellation of Capricornus is perhaps the first constellation to have been recognized, probably in prehistoric times. Depictions of a goat, or of a goat-fish, have been found on Babylonian tablets around 3,000 years old. Ancients associated the constellation with Amalthea, the goat that suckled the infant Zeus, and also with the god Pan. About 2,000 years ago the Sun reached its most southerly position in the sky—the winter solstice (23.4° S latitude)—when it passed in front of this constellation. This latitude thus came to be called the Tropic of Capricorn; it is the most southerly latitude at which the Sun is ever directly overhead. However, the Sun is no longer in Capricorn at the winter solstice because of the change in the orientation of Earth's axis (precession).

Stars None notable; the brightest is of only 3rd magnitude.

Deep-Sky Objects M30 is a fairly compact globular star cluster of 7th magnitude, located about 4° east-southeast of ζ Capricorni.

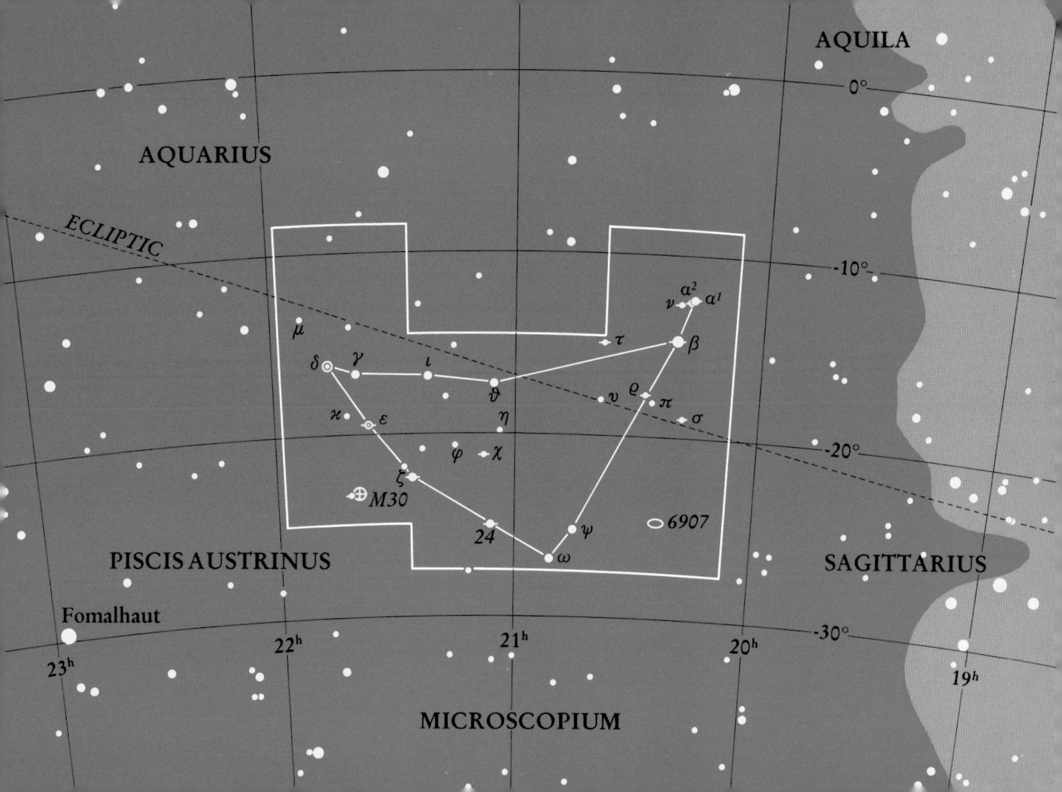
AQUILA
0°
AQUARIUS
ECLIPTIC
-10°
α²
ν
α¹
μ
τ
β
δ
γ
ι
ϑ
υ
ϱ
π
ϰ
ε
η
σ
φ
χ
-20°
ζ
M30
6907
24
ψ
ω
PISCIS AUSTRINUS
SAGITTARIUS
Fomalhaut
-30°
22h
21h
20h
23h
19h
MICROSCOPIUM

Cassiopeia (Cas) The Queen

On Meridian: November 20   Area: 598 square degrees

Cassiopeia is a bright configuration of stars in the form of a W (or M) located along a bright part of the Milky Way. Cassiopeia and neighboring Cepheus were the parents of Princess Andromeda, also nearby.

Stars

α, named Schedar (Arabic for "beast"), is a slightly variable star. Spectral type K0 II–III; magnitude 2.2; distance 120 ly. γ, Cih ("whip" in Chinese), is a B0 IV subgiant star about 730 ly distant. Its magnitude varies from 1.6 to 3.0 in an unpredictable manner. η, sometimes called Achird, lies beside the line between Schedar (α Cas) and Cih (γ Cas). Spectral type G0 V; magnitude 3.4; distance 19 ly. η is an attractive double star, with a type-K companion of 7th magnitude.

Deep-Sky Objects

There is a wealth of clusters and nebulae visible with binoculars or a telescope. M52, a good example of an open cluster, contains more than 100 stars, the brightest of which is around 7th magnitude, within an area about 12′ across. M103 is a smaller open cluster with fewer stars.

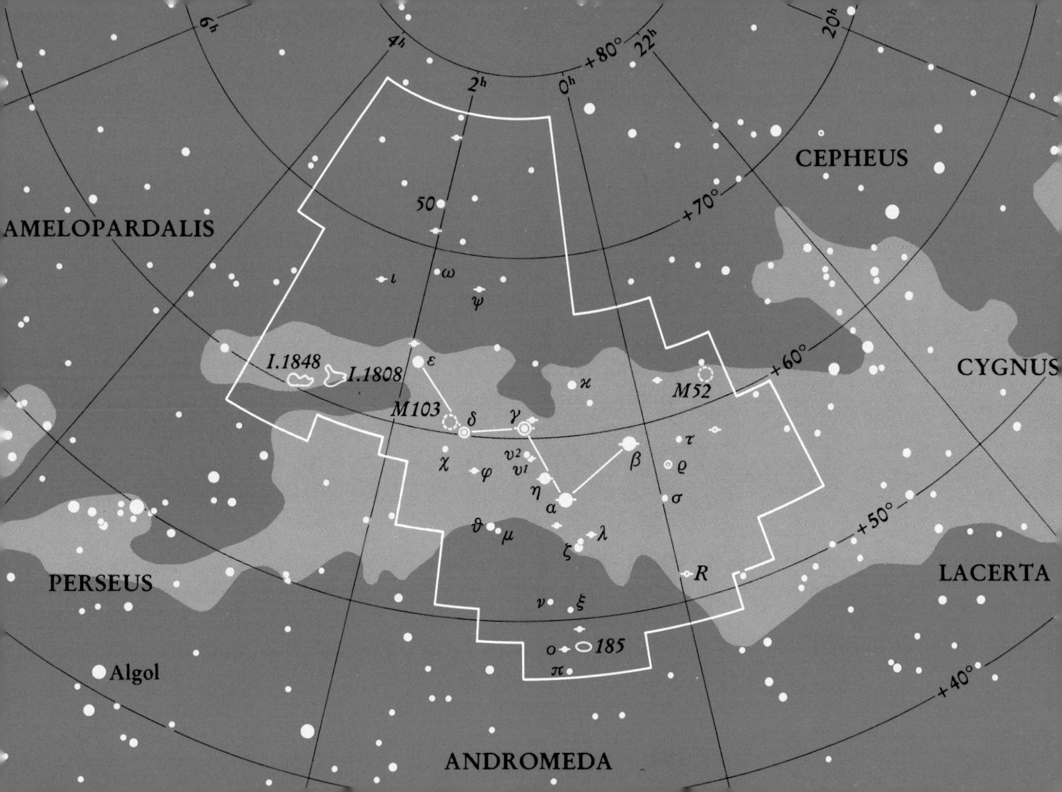

6h
4h
2h
0h
22h
20h
+80°
+70°
+60°
+50°
+40°
CEPHEUS
AMELOPARDALIS
CYGNUS
LACERTA
PERSEUS
ANDROMEDA
Algol
50
ι
ω
ψ
ε
I.1848
I.1808
M103
δ
γ
ϰ
M52
τ
β
ϱ
σ
χ
φ
υ2
υ1
η
α
ϑ
μ
λ
ζ
R
ν
ξ
o
185
π

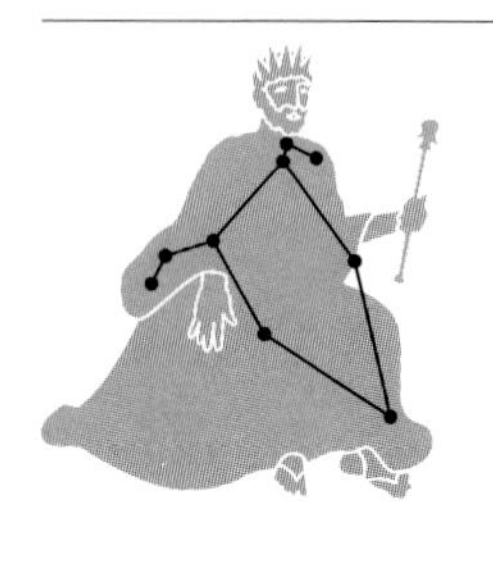

Cepheus (Cep) The King

On Meridian: October 15   Area: 588 square degrees

Although it has no very bright stars, Cepheus has been well known for thousands of years. Its mythological namesake was the husband of Cassiopeia and the father of Andromeda. Its shape is rather like a child's drawing of a house with a steep roof. Because the Milky Way passes through Cepheus it is rich in star clusters and nebulae.

Stars

α is called Alderamin, which in Arabic means either "forearm" or "shoulder." Spectral type A7 IV–V; magnitude 2.4; distance 49 ly. δ is the prototype of the important class of variable stars called Cepheids. Over a period of precisely 5.366341 days δ Cephei varies from magnitude 3.5, spectral type F5, to magnitude 4.4, spectral type G2, and back again.

Deep-Sky Objects

NGC 188, the oldest open cluster known, is of 9th magnitude overall, but its brightest stars are of only about 11th magnitude. IC 1396, southwest of μ Cephei, contains about 30 stars in a region some 50′ in diameter, almost twice the size of the full moon. Its brightest stars are of about 6th magnitude.

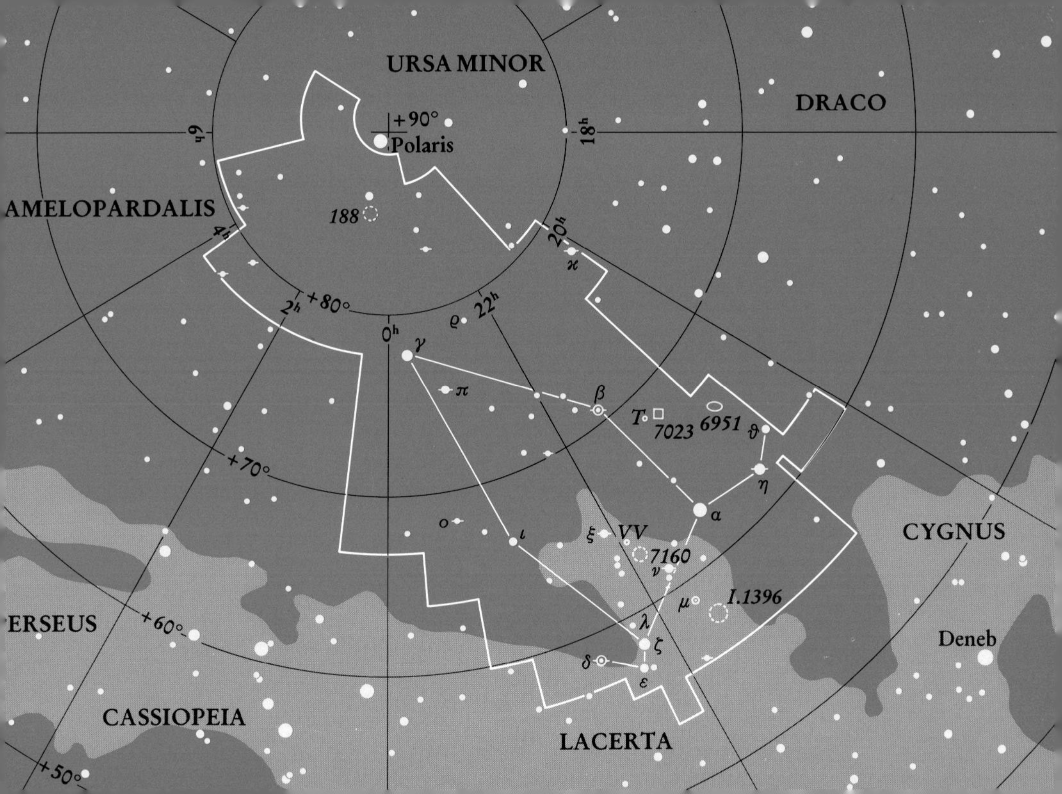
URSA MINOR
DRACO
+90°
Polaris
AMELOPARDALIS
188
6h
18h
4h
20h
2h
+80°
0h
22h
γ
π
β
T
7023
6951
ϑ
η
α
+70°
ι
ξ
VV
7160
ν
μ
I.1396
λ
ζ
δ
ε
CYGNUS
Deneb
ERSEUS
+60°
CASSIOPEIA
LACERTA
+50°

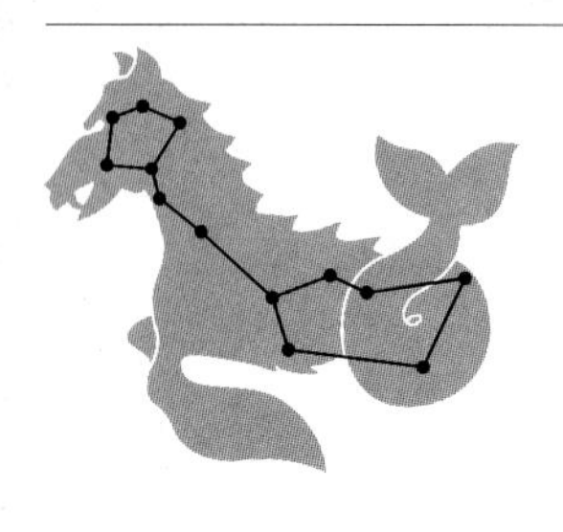

Cetus (Cet) The Sea Monster, the Whale
On Meridian: November 30   Area: 1,231 square degrees

Cetus is among the earliest constellations identified and the fourth largest, but its brightest star is of only 2nd magnitude. Ancient Mesopotamian civilizations identified these stars as Tiamat, the cosmic dragon slain by the hero Marduk. In classical mythology Cetus is the sea monster that threatened Andromeda.

Stars

o is Mira, the prototype of the class of pulsating variables called long-period variables (LPVs) and the first variable star discovered, in 1596. Its discovery lent support to the Copernican revolution in astronomy and furthered the collapse of the old beliefs, which held that the sky never changed. Over an average period of 330 days Mira goes from spectral type M5 III, magnitude 3.4 (it has reached 2nd magnitude), to spectral type M9 III, magnitude 9.3, and back again. It is 130 ly away.

Deep-Sky Objects

Cetus contains a wealth of faint galaxies, the brightest of which is M77, a 9th-magnitude spiral galaxy.

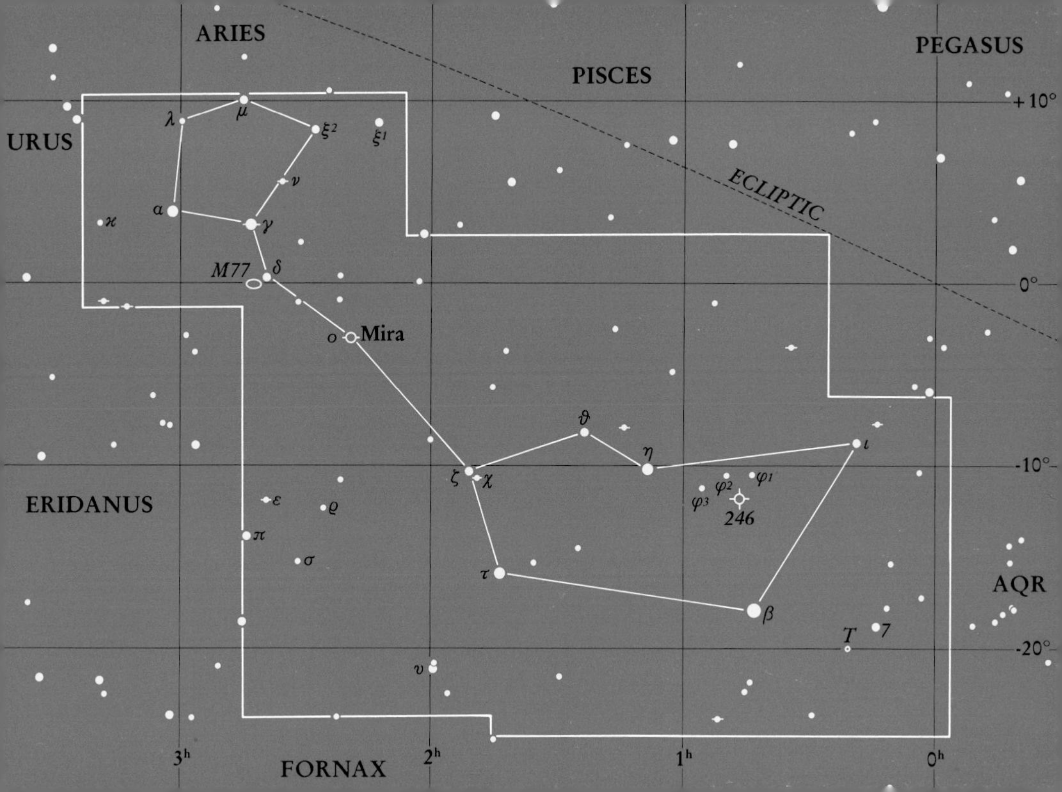

ARIES
PISCES
PEGASUS
URUS
ECLIPTIC
M77
Mira
ERIDANUS
AQR
246
T
7
+10°
0°
-10°
-20°
3h
2h
1h
0h
FORNAX

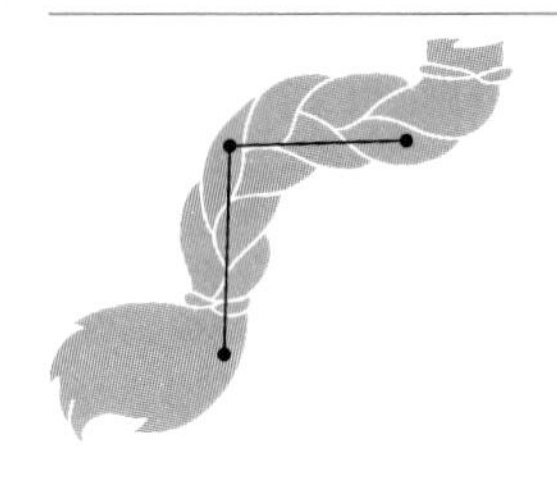

Coma Berenices (Com) Berenice's Hair
On Meridian: May 15   Area: 386 square degrees

Although the configuration of this small, dim constellation was not precisely drawn until 1602, Coma Berenices is of ancient origin and represents the hair of Berenice II, queen of Egyptian ruler Ptolemy III Euergetes (fl. 246–221 B.C.). These stars had once been considered the tuft at the end of the tail of the neighboring constellation Leo. When you look toward Coma Berenices you are looking north, perpendicular to the plane of the Milky Way, and hence have a clear view of space beyond our own galaxy.

Stars

β is just slightly more luminous than the Sun, giving us an idea of how dim our own star would appear from the cosmically close distance of 27 ly. Spectral type G0 V; magnitude 4.6.

Deep-Sky Objects

The Coma galaxy cluster, millions of light-years away, contains more than 1,000 galaxies, most fainter than 12th magnitude. Some closer galaxies can be seen with a small telescope near the Virgo border. M64 (the Black-Eye Galaxy) and M85 are both 9th-magnitude spirals.

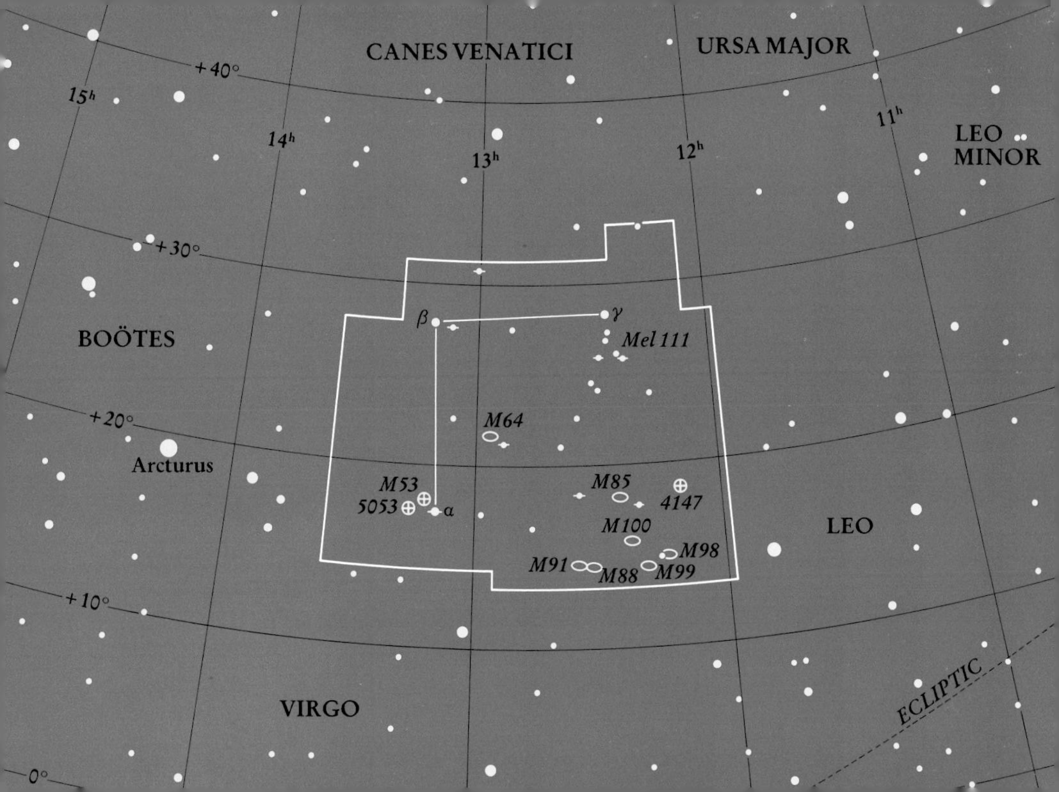

CANES VENATICI
URSA MAJOR
LEO
MINOR
+40°
15h
14h
13h
12h
11h
+30°
+20°
+10°
0°
BOÖTES
Arcturus
β
γ
α
Mel 111
M64
M53
5053
M85
4147
M100
M98
M91
M88
M99
LEO
VIRGO
ECLIPTIC

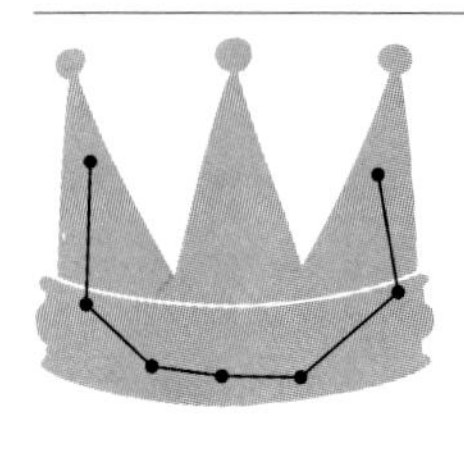

Corona Borealis (CrB) The Northern Crown
On Meridian: June 30   Area: 179 square degrees

Corona Borealis is among the smaller constellations, yet it is renowned for its beautiful shape, which resembles a crown. This incomplete circlet of 3rd- and 4th-magnitude stars has been recognized by many different civilizations worldwide. In classical myth it is the crown of Ariadne, daughter of the king of Crete. To the Native American Shawnee tribe it was a circle of dancing star maidens.

Stars

α, called Gemma, the gem, or Alphecca, Arabic for "the brightest," is really a pair of stars of a type known as an eclipsing binary. The two stars are of types A0 V and G5 V but are so close together they appear as one star of magnitude 2.2 that dims slightly during the eclipses. They lie 78 ly away.

Deep-Sky Objects

Although there are more than 400 galaxies clustered in the southwestern corner of Corona Borealis, none is visible in even moderate-size telescopes, as the brightest is of only 16th magnitude. They are estimated to be around a billion light-years from our own galaxy.

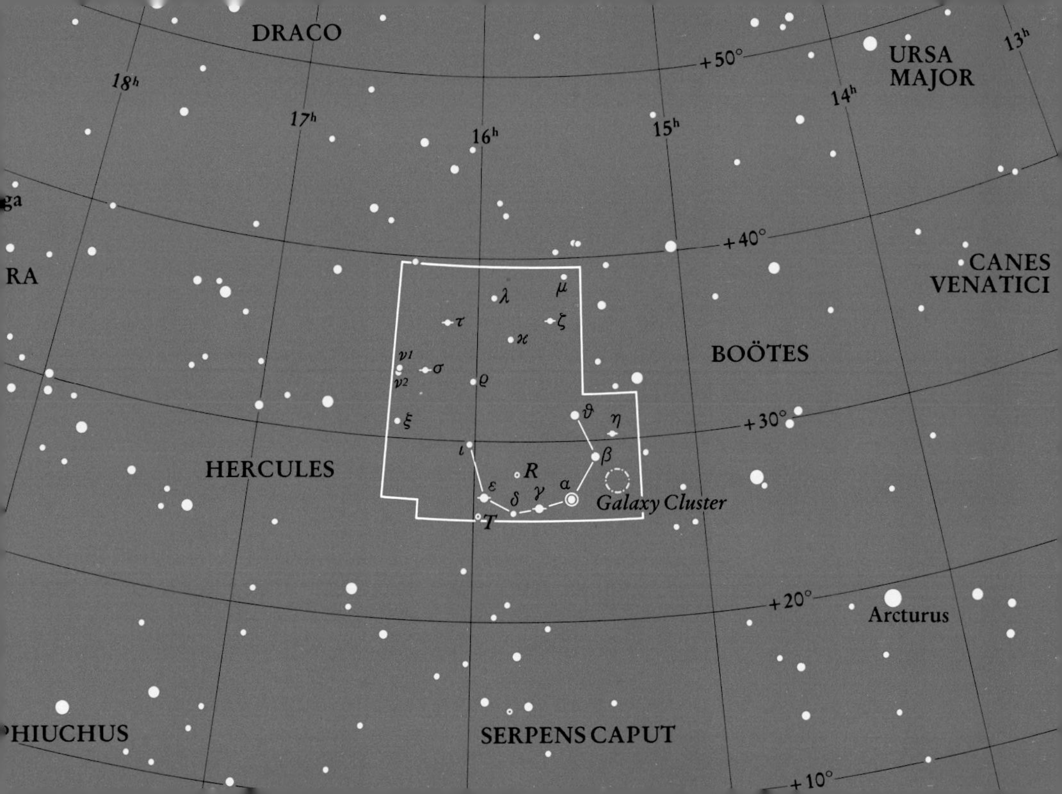

DRACO
+50°
URSA MAJOR
13h
14h
15h
16h
17h
18h
ga
RA
+40°
CANES VENATICI
λ
μ
τ
ζ
ϰ
ν1
σ
ν2
ϱ
BOÖTES
ϑ
η
ξ
+30°
ι
β
HERCULES
R
ε
α
δ
γ
Galaxy Cluster
T
+20°
Arcturus
PHIUCHUS
SERPENS CAPUT
+10°

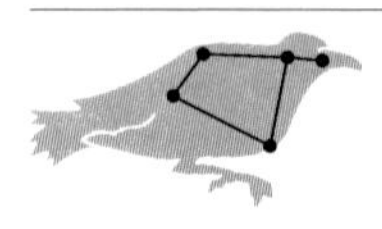

Corvus (Crv) The Crow, the Raven
On Meridian: May 10   Area: 184 square degrees

According to legend, Corvus, identified as a crow or a raven, was sent to fetch a cup of water for Apollo. He dawdled and was banished to the sky, where he sits within sight of the cup, Crater, but cannot drink. The four main stars of Corvus, forming a trapezium, are quite distinctive and can be found below the stars of Virgo, southwest of the bright star Spica. Corvus has several galaxies, none brighter than 11th magnitude.

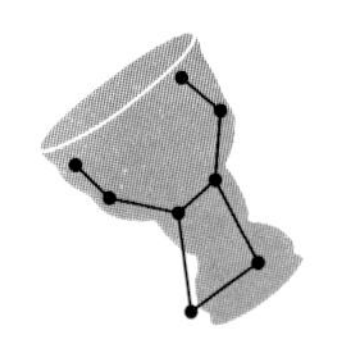

Crater (Crt) The Cup
On Meridian: April 25   Area: 282 square degrees

Crater, the cup carried by Corvus the crow to Apollo, is also sometimes said to be the cup of nectar drunk by the Olympian gods. About 25° south of Denebola, the β star in Leo, Crater is small and consists of no bright stars. With a certain amount of imagination you may be able to see a cup or goblet here. There are a number of galaxies in Crater, all too faint for amateur instruments.

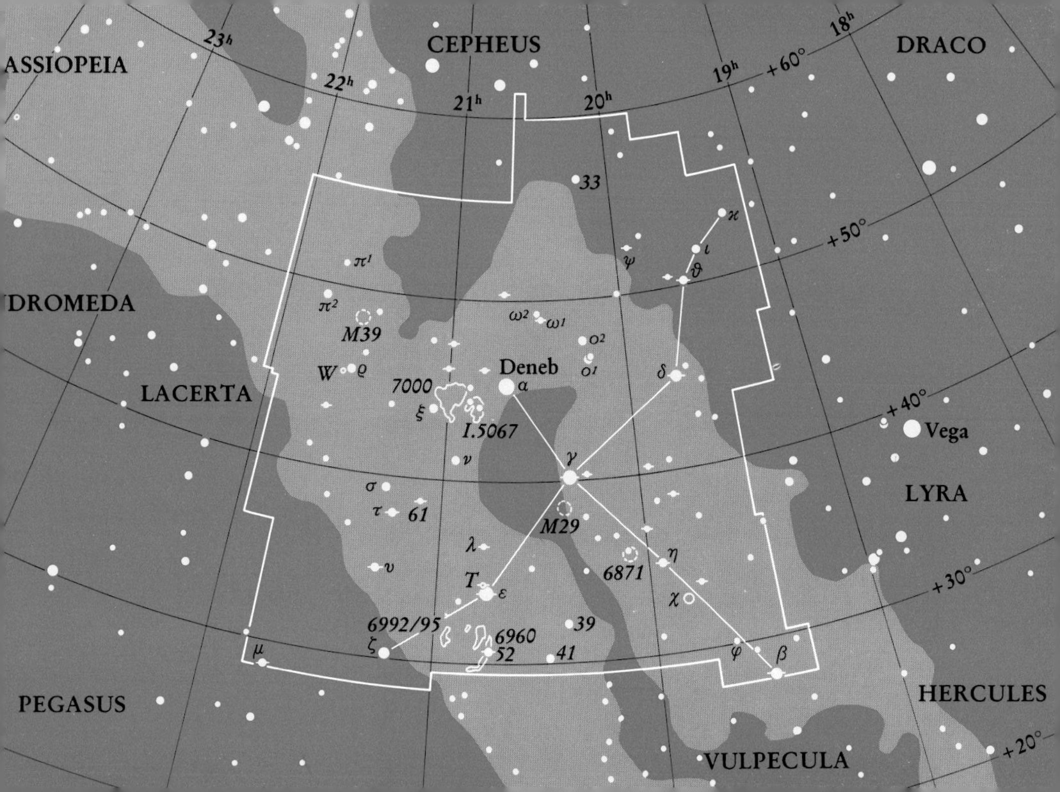

CEPHEUS
DRACO
ASSIOPEIA
23h
22h
21h
20h
19h
18h
+60°
+50°
+40°
+30°
+20°
DROMEDA
LACERTA
PEGASUS
LYRA
Vega
HERCULES
VULPECULA
Deneb
α
γ
δ
ε
ζ
η
θ
ι
κ
λ
μ
ν
ξ
ρ
σ
τ
υ
φ
χ
ψ
β
π1
π2
ω1
ω2
o1
o2
W
T
M39
M29
7000
I.5067
6871
6992/95
6960
61
52
41
39
33

Delphinus (Del) The Dolphin

On Meridian: September 15   Area: 189 square degrees

Delphinus is among the smaller constellations, but it has a distinctive shape that has long been recognized as a dolphin. In every ancient myth concerning the animal, the dolphin is seen as a friend and rescuer of gods and men. To find Delphinus, look about 30° south of α Cygnus (Deneb) and between the bright α star in Aquila (Altair), and the Great Square of Pegasus. With a little effort you can imagine the arched body of a dolphin, or a tiny kite with a tail.

Stars — α was named Sualocin and β was named Rotanev in an 1814 star catalog published by the Palermo Observatory. Read backward the two names spell "Nicolaus Venator," the Latinized form of the Italian name Niccolo Cacciatore, the assistant to the director of the observatory, who bestowed this honor upon him. α: spectral type B9 V; magnitude 3.8; distance 170 ly. β: spectral type F5 III; magnitude 3.8; distance 110 ly.

Deep-Sky Objects — NGC 6934 is a 9th-magnitude globular cluster about 1.5′ across, 54,000 ly from Earth.

CYGNUS
+30°
HERCULES
VULPECULA
+20°
PEGASUS
6905
7006
SAGITTA
α
γ
δ
ζ
β
η
OPH
ι
ε
+10°
ϰ
Altair
EQUULEUS
6934
SERPENS
CAUDA
AQUILA
0°
22h
19h
21h
20h
SCUTUM
AQUARIUS
ECLIPTIC
-10°

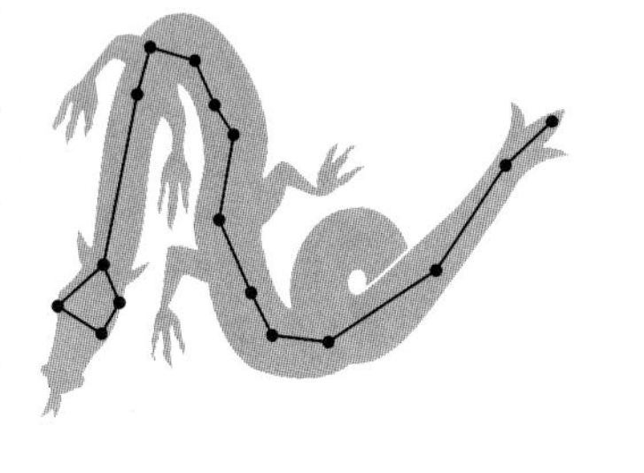

Draco (Dra) The Dragon
On Meridian: July 20   Area: 1,083 square degrees

Draco has stood for all the dragons of mythology, from Tiamat of the Sumerians to the monster slain by Saint George. The sinuous pattern of its stars is certainly dragonlike. It winds around the sky close to the north celestial pole. Because the Earth's axis—and by extension the celestial poles—shifts position over time (a process called precession), the star closest to the north celestial pole also changes. Today Polaris (the α star in Ursa Minor) is the polestar, but 4,000 years ago that position was occupied by Draco's α star, Thuban.

Stars — α, Thuban, is unremarkable now, but when it was the polestar Egyptians oriented temples to it. Spectral type A0 III; magnitude 3.7; distance 230 ly. β is named Rastaban, "head of the dragon." Spectral type G2 II; magnitude 2.8; distance 490 ly.

Deep-Sky Objects — NGC 6543 is a 9th-magnitude planetary nebula about 22″ across. There are a number of galaxies in Draco, most fainter than 10th magnitude.

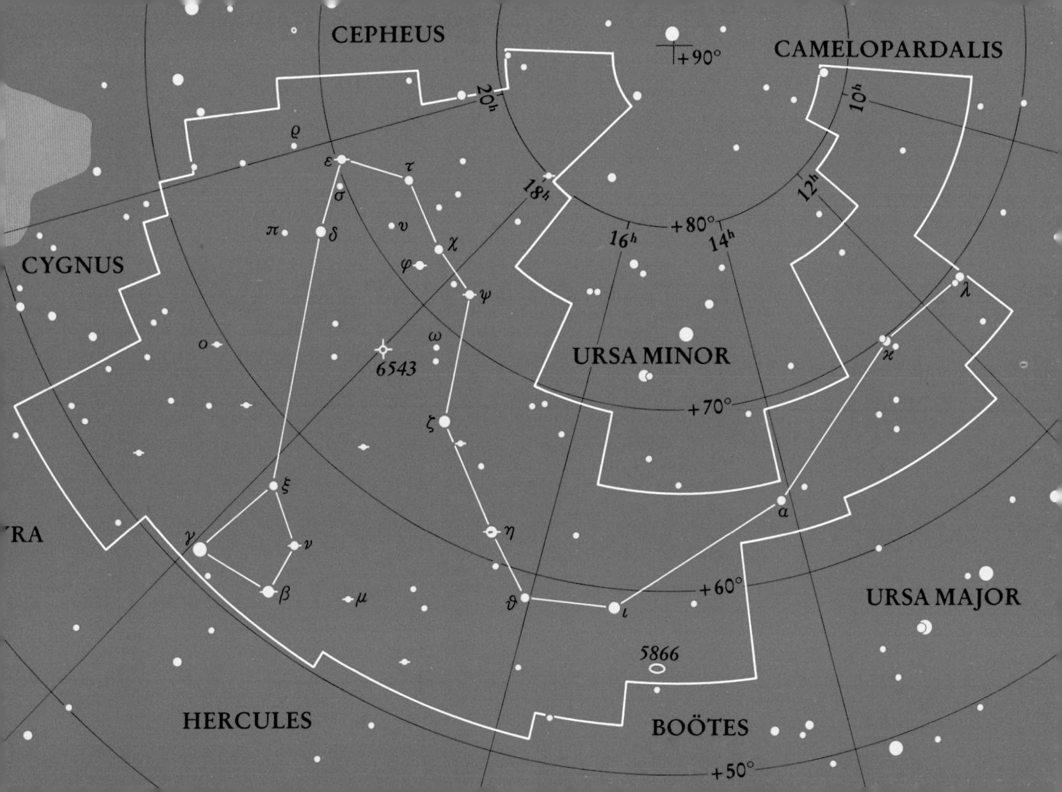
CEPHEUS
+90°
CAMELOPARDALIS
20h
10h
18h
12h
+80°
16h
14h
CYGNUS
URSA MINOR
6543
+70°
'RA
+60°
URSA MAJOR
5866
HERCULES
BOÖTES
+50°

Equuleus (Equ) The Little Horse, the Colt
On Meridian: September 20   Area: 72 square degrees

This constellation is said to have been named by the Greek astronomer Hipparchus in the 2nd century B.C., but for many centuries it was rarely mentioned in astronomy works. It is probably referred to as the "Little" Horse to distinguish it from nearby Pegasus, the Winged Horse. Equuleus lies between the "nose" of Pegasus, to the east, and the small constellation of Delphinus, to the west. It is the second-smallest constellation in area (only Crux, the Southern Cross, is smaller, by just 4 square degrees). Because Equuleus is small and has no bright stars, it is difficult to find and of comparatively little interest.

Stars — α is named Kitalpha, Arabic for "part of a horse" or "little horse." Spectral type G0 III; magnitude 3.9; distance 150 ly.

Deep-Sky Objects — None.

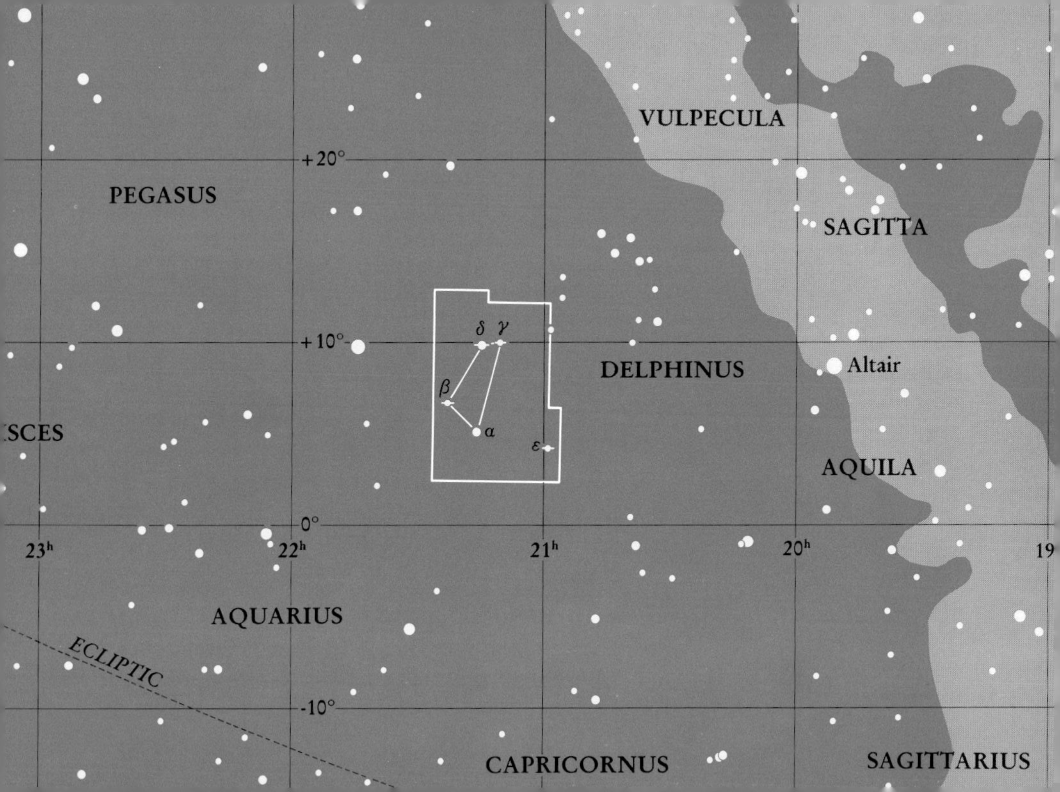

VULPECULA
+20°
PEGASUS
SAGITTA
δ
γ
+10°
DELPHINUS
Altair
β
SCES
α
ε
AQUILA
0°
23h
22h
21h
20h
19
AQUARIUS
ECLIPTIC
-10°
CAPRICORNUS
SAGITTARIUS

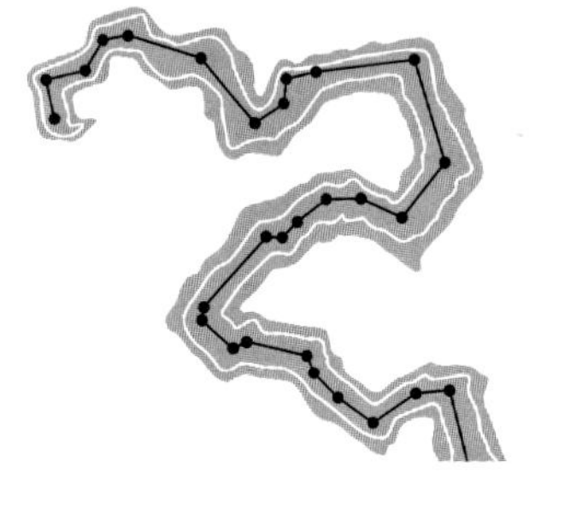

Eridanus (Eri) The River, the Celestial River
On Meridian: January 5   Area: 1,138 square degrees

Eridanus is the second-longest (after Hydra), but only the sixth-largest, constellation in the sky. These faint stars have been known as a river (usually thought of as the Euphrates or the Nile) since ancient times. The constellation winds westward toward Cetus, then southward between Fornax and Caelum, a constellation of the southern skies. In classical times the southernmost limit of the constellation was the star Acamar, ϑ Eridani. It was later extended southward to the star Achernar, α Eridani, a 1st-magnitude star; our map does not show the far southern reaches of Eridanus.

Stars — α is Achernar, from the Arabic word for “river’s end.” It is visible only south of latitude 30° N (not on map). Spectral type B3 V; magnitude 0.5; distance 69 ly.

Deep-Sky Objects — IC 2118, a very large (about $140' \times 40'$) reflection nebula just south of β Eridani, is 2.5° northwest of Rigel (β Ori) and presumably illuminated by that brilliant star. There are several dozen faint galaxies (10th magnitude or less) strewn throughout Eridanus.

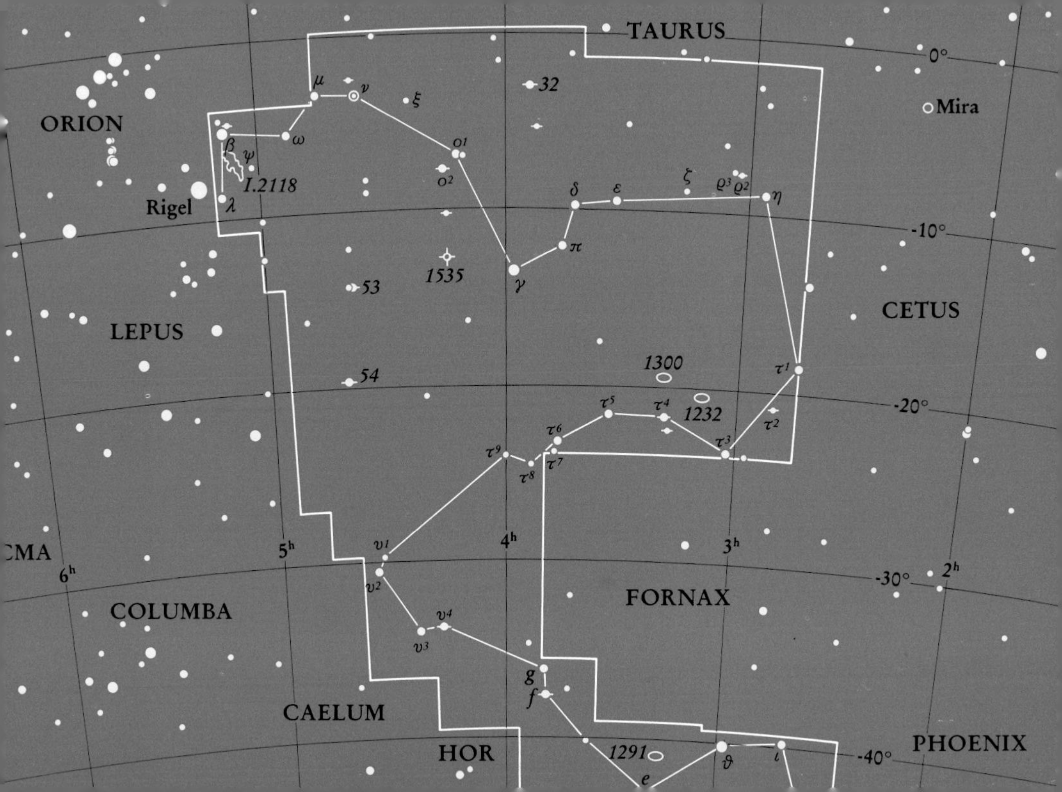
TAURUS
0°
32
Mira
ORION
μ
ν
ξ
ω
ο1
β
ψ
I.2118
ο2
ζ
ϱ3 ϱ2
δ
ε
η
Rigel
λ
-10°
π
1535
53
γ
CETUS
LEPUS
1300
τ1
54
τ5
τ4
1232
-20°
τ6
τ3
τ2
τ9
τ8
τ7
CMA
5h
4h
3h
υ1
6h
υ2
-30°
2h
FORNAX
COLUMBA
υ4
υ3
g
f
CAELUM
HOR
1291
PHOENIX
-40°
ϑ
ι
e

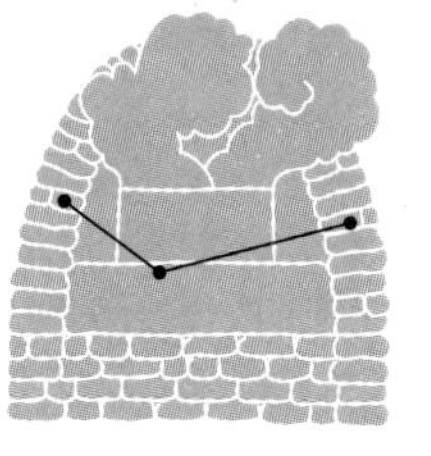

Fornax (For) The Furnace

On Meridian: December 15   Area: 398 square degrees

The originator of this constellation, the 18th-century French astronomer Nicolas-Louis de Lacaille, called it Fornax Chemica, the Chemical Furnace. He composed it from some faint stars lying within one of the bends of the river Eridanus. Fornax contains no bright stars and only faint deep-sky objects.

Stars — α: spectral type F8 IV; magnitude 3.9; distance 46 ly.

Deep-Sky Objects — None visible with a small telescope, although this is the home of the Fornax galaxy cluster, which lies in the southeastern corner of Fornax and spills across its border into Eridanus, covering about 1 square degree. It consists of about 18 closely grouped elliptical and spiral galaxies, ranging in magnitude from 10 to 13.

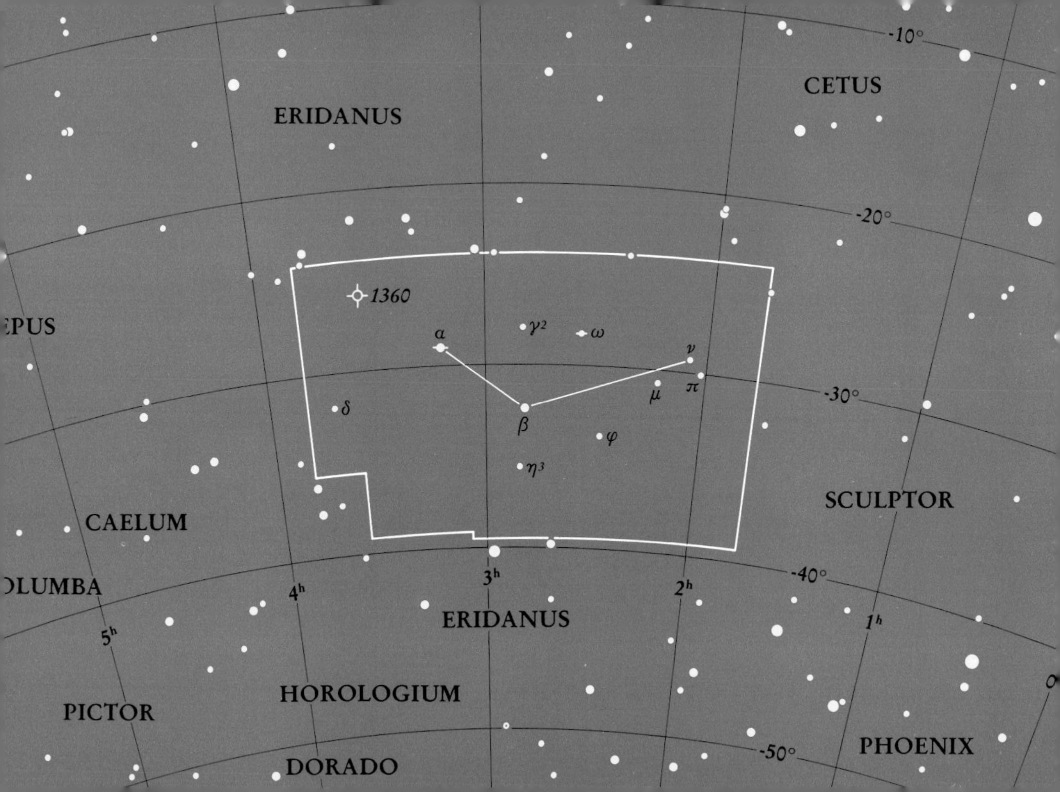

-10°
CETUS
ERIDANUS
-20°
1360
EPUS
α
γ2
ω
ν
π
μ
-30°
δ
β
φ
η3
SCULPTOR
CAELUM
-40°
OLUMBA
4h
3h
2h
ERIDANUS
1h
5h
HOROLOGIUM
0
PICTOR
-50°
PHOENIX
DORADO

Gemini (Gem) The Twins

On Meridian: February 20   *Area: 514 square degrees*

Gemini is dominated by its two bright stars, Castor and Pollux, the sons of Leda and Zeus (disguised as a swan) and the brothers of Helen of Troy. Part of the zodiac, Gemini is a familiar constellation of the winter sky.

Stars

α, Castor, is Gemini's second-brightest star. To the naked eye it appears as a single star of spectral type A1 V and magnitude 1.6, but it is actually a sextuple. In a telescope it looks like two stars of magnitudes 2 and 3, only 2″ apart, with a 9th-magnitude companion 73″ from them. Each of these three stars is itself double. The system of stars is 46 ly from Earth. β, Pollux, is the brightest star of Gemini. Spectral type K0 III; magnitude 1.1; distance 36 ly.

Deep-Sky Objects

M35 (NGC 2168) is a bright open cluster of about 5th magnitude, 2,800 ly away. It consists of approximately 120 stars in a region 40′ across and is located near the western edge of Gemini, northwest of η Geminorum. NGC 2392, near δ Geminorum, is an 8th-magnitude planetary nebula about 13″ in diameter.

LYNX
AURIGA
9h
8h
7h
6h
5h
+40°
+30°
+20°
+10°
Castor
Pollux
TAURUS
EO
ECLIPTIC
2158
M35
I.2157
I.444
2129
2392
CANCER
ORION
Betelgeuse
Procyon
IYDRA
CANIS MINOR
MONOCEROS

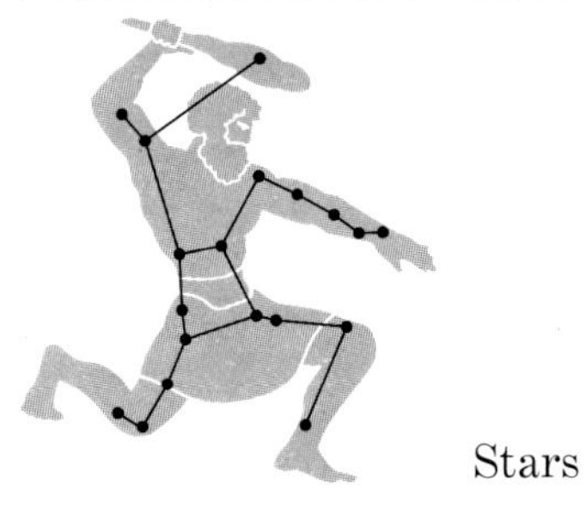

Hercules (Her) The Strongman
On Meridian: July 25   Area: 1,225 square degrees

Hercules, one of the earliest constellations named, is probably connected with earlier strongman-heroes, such as the Sumerian Gilgamesh. It takes a bit of imagination to see a man (upside down), kneeling, with his foot on the head of Draco, the dragon, in these dim stars.

Stars

α is named Ras Algethi, Arabic for "the kneeler's head." A 3rd-magnitude star to the naked eye, it is revealed in a small telescope to be a beautiful double star. The brighter is an M5 II giant, distinctly red in color, while its companion is a 5th-magnitude G5 III star, greenish by contrast. The stars are 630 ly away.

Deep-Sky Objects

M13 is the most spectacular globular cluster visible from the Northern Hemisphere. On clear nights it appears to the naked eye as a 6th-magnitude, slightly fuzzy star along the line marking the western edge of the figure. It measures about 10′ in diameter, contains perhaps 300,000 stars, and lies some 23,000 ly from Earth. M92, another globular cluster, is almost at bright as M13.

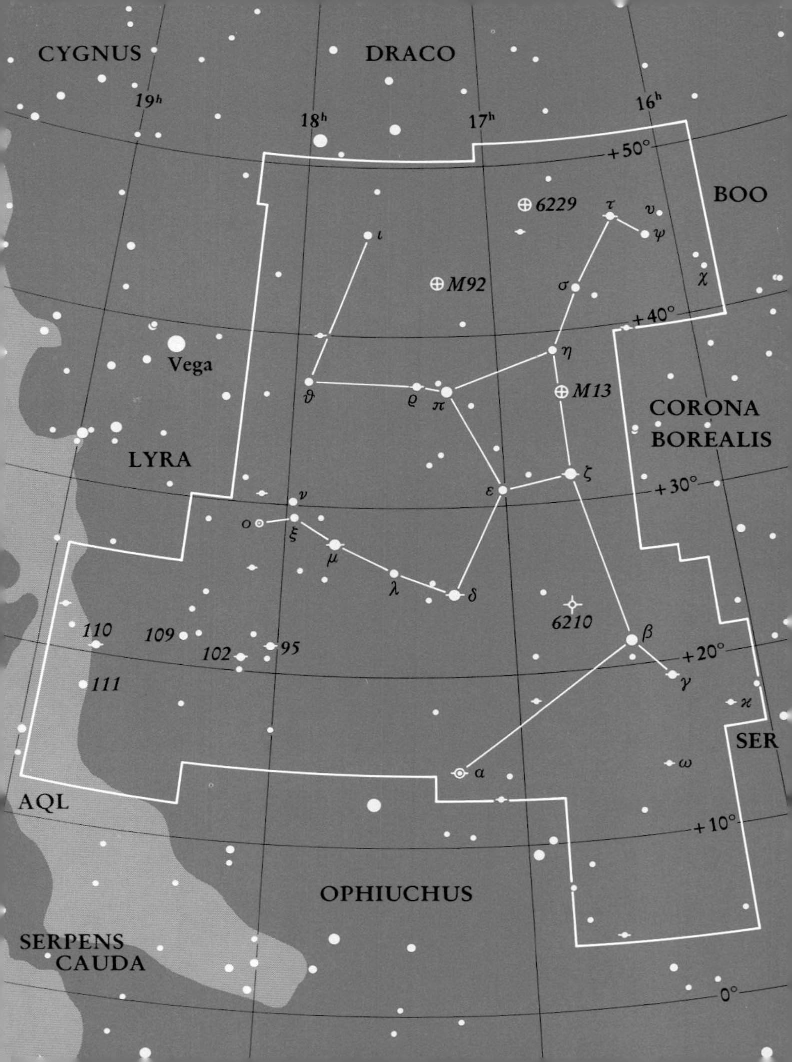

CYGNUS
DRACO
19h
18h
17h
16h
+50°
+40°
+30°
+20°
+10°
0°
BOO
6229
M92
M13
Vega
LYRA
CORONA
BOREALIS
6210
110
109
102
95
111
SER
AQL
OPHIUCHUS
SERPENS
CAUDA

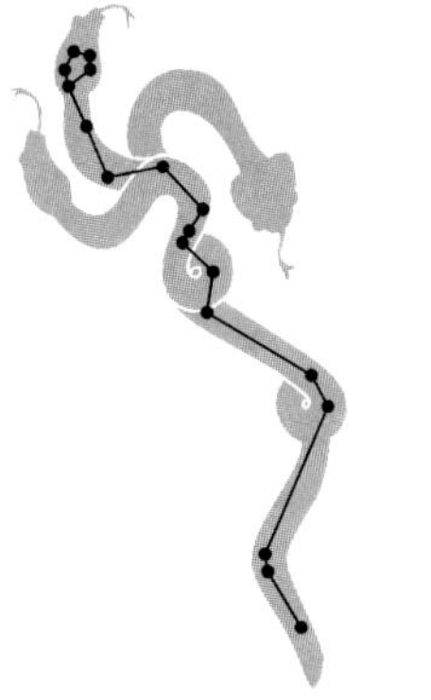

Hydra (Hya) The Sea Serpent
On Meridian: April 20   Area: 1,303 square degrees

Hydra is the largest and longest constellation in the sky. At mid-northern latitudes, it takes more than six hours for the whole constellation to rise. Hydra was the many-headed monster slain by Hercules. (It is not to be confused with the modern constellation of Hydrus, the Water Snake, composed by Johann Bayer near the south celestial pole in his 1603 atlas.) Hydra's head is south of Cancer and east of Canis Minor. The monster's sinuous body snakes eastward almost as far as Scorpius, a quarter of the way around the sky.

Stars α is Alphard (or Alfard), which comes from Arabic and means "the solitary one." It is the only named star in Hydra and by far the brightest. Spectral type K3 III; magnitude 2.0; distance 110 ly. ϵ appears as a G5 III star of magnitude 3.4 but is actually a quadruple, with components of 3rd, 5th, 8th, and 12th magnitudes. The system lies 150 ly from Earth.

Deep-Sky Objects M48, located on Hydra's western border with Monoceros, is a large, bright open cluster of about 80 stars covering a region about the same angular size as the full moon. The cluster lies 3,100 ly away and is a bit brighter than 6th magnitude. M68 is a globular cluster measuring about 9′ across, located just south of Hydra's border with Corvus. The cluster is of 8th magnitude, but the brightest individual stars are of only 13th magnitude. M83, an 8th-magnitude spiral galaxy, is among the brightest galaxies in the southern skies. Because of its southerly position, it is difficult for northern observers to see. It lies about 18° south of the star Spica (α Vir), near the tail end of Hydra. There are dozens of other galaxies in Hydra, but all are of 11th magnitude or fainter.

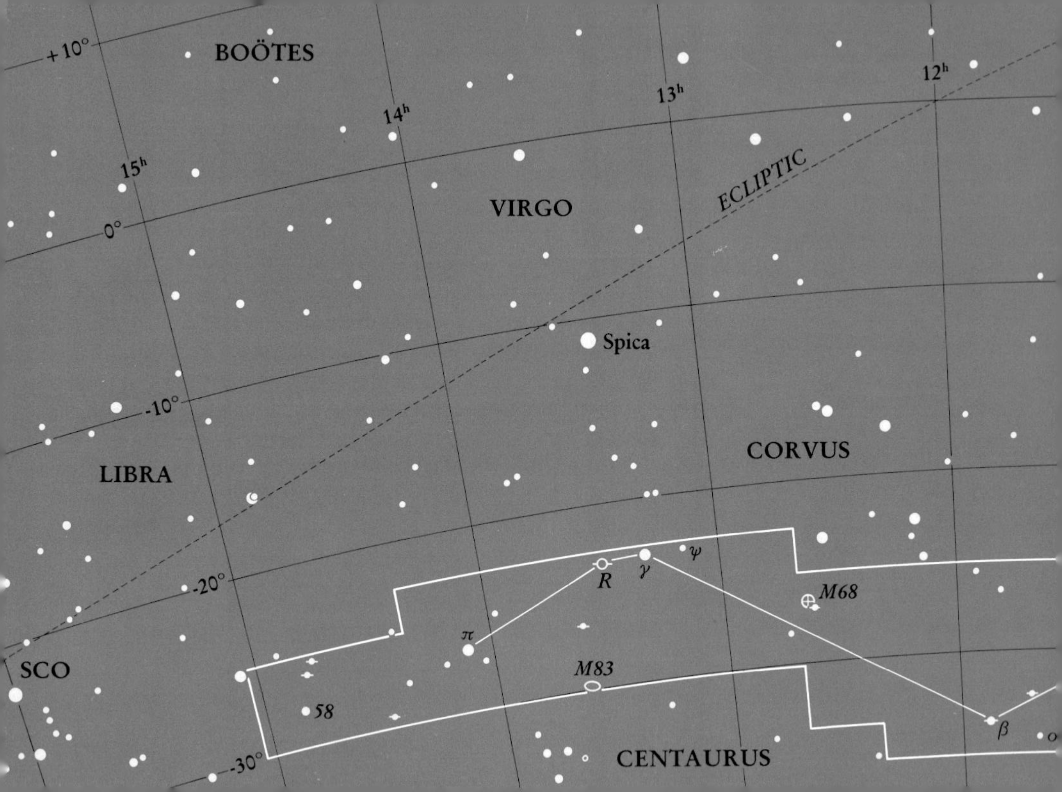

BOÖTES
+10°
12h
13h
14h
15h
ECLIPTIC
VIRGO
0°
Spica
-10°
CORVUS
LIBRA
-20°
R
γ
ψ
M68
π
SCO
M83
58
β
-30°
CENTAURUS

CANCER
+10°
11h
10h
9h
8h
ε
ω
ζ
ϱ
δ
ϑ
η
σ
Procyon
CANIS MINOR
ι
τ2
τ1
0°
SEXTANS
C
M48
α
λ
υ2
U
MONOCEROS
ν
υ1
ϰ
-10°
12
φ
μ
3242
-20°
CANIS MAJOR
Sirius
χ
PUPPIS
ANTLIA
PYXIS
-30°

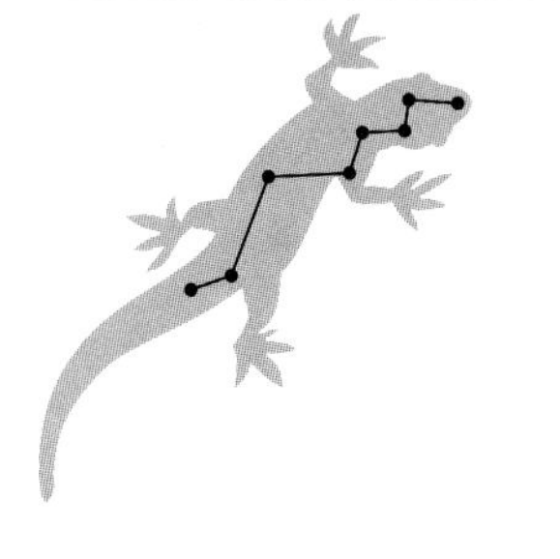

Lacerta (Lac) The Lizard

On Meridian: October 10   Area: 201 square degrees

Lacerta is a modern constellation, created around 1687 by the German astronomer Johannes Hevelius to fill in an unnamed region between Cygnus and Andromeda. He described it as a lizard or a newt. Despite its small size, Lacerta's location on the edge of the band of the Milky Way means that it contains several deep-sky objects, mostly open clusters.

Stars — All of 4th magnitude or fainter.

Deep-Sky Objects — NGC 7209, located on the western edge of Lacerta on its border with Cygnus, is an open cluster of about 50 stars with magnitudes ranging from 9 to 12. It covers a region about 20′ in diameter and lies 2,900 ly away. NGC 7243, an open cluster of about 8th magnitude, has about 40 stars. Located west of α Lacertae, it is 20′ in diameter and 2,800 ly distant.

CEPHEUS
2h
1h
0h
23h
22h
21h
20h
19h
+60°
+50°
+40°
+30°
+20°
RSEUS
CASSIOPEIA
7296
β
I.5217
α
4
7243
5
7209
2
11
6
1
Deneb
CYGNUS
ANDROMEDA
SCES
PEGASUS
VULPECULA

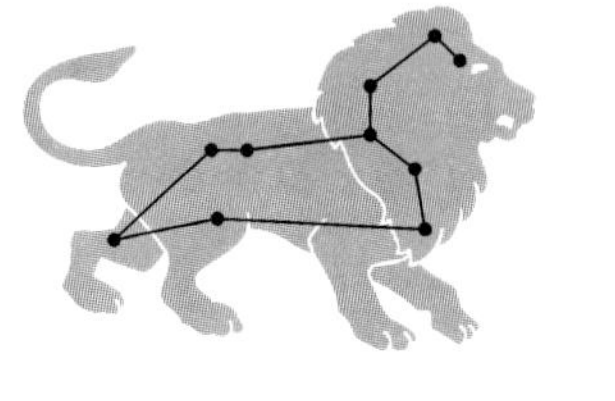

Leo (Leo) The Lion

On Meridian: April 10   Area: 947 square degrees

The stars we call Leo were recognized as a lion by the ancient Sumerians, Babylonians, Persians, Syrians, Greeks, and Romans. They were seen as a horse in the ancient Chinese zodiac, and possibly as a puma in Incan lore. Leo's head and mane are formed by an asterism known as the Sickle (sometimes viewed as a backward question mark or fishhook). The hindquarters are formed by a triangle of stars.

Stars α, Regulus (Latin for "little king"), is the faintest of the 1st-magnitude stars, with a magnitude of 1.4. Its spectral type is B7 V, and it is 69 ly away. It forms the dot below the backward question mark (the handle of the Sickle). It lies very close to the ecliptic. β is called Denebola, Arabic for "lion's tail." Spectral type A3 V; magnitude 2.1; distance 39 ly.

Deep-Sky Objects There are many galaxies in Leo, but the brightest are only of 9th magnitude: M65 and M66 (a close pair), NGC 3628, M95 and M96 (also close), all spirals.

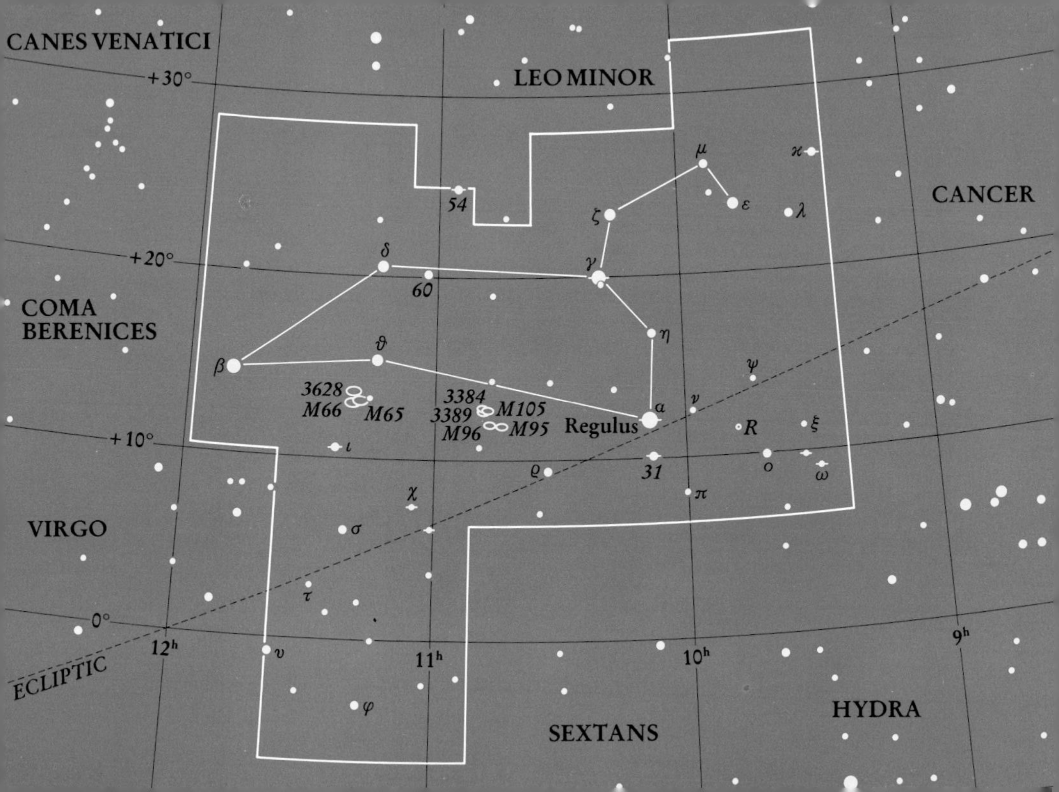

CANES VENATICI
LEO MINOR
CANCER
COMA BERENICES
VIRGO
HYDRA
SEXTANS
ECLIPTIC
+30°
+20°
+10°
0°
12h
11h
10h
9h
Regulus
3628
M66
M65
3384
3389
M105
M96
M95
54
60
31
R

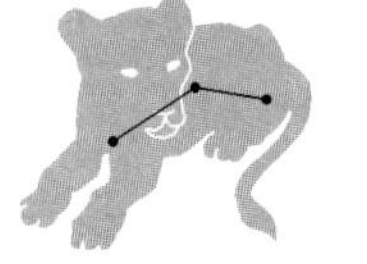

Leo Minor (LMi) The Little Lion
On Meridian: April 10   Area: 232 square degrees

This is one of several constellations named in modern times (circa 1687) to fill in areas of the northern sky known to the ancient Greeks as *amorphotoi*, the "unformed" or "unshaped." It is thought that these stars may have represented a gazelle to the ancient Arabs. In Chinese lore they were sometimes combined with the stars of Leo to make a huge celestial dragon and, in another depiction, a chariot. Leo Minor's faint stars lie between two more conspicuous constellations: Leo, to the south, and Ursa Major, to the north. Only one star in Leo Minor (β) retains its Greek-letter designation.

Stars — 46 Leonis Minoris is the brightest star of the constellation, at magnitude 3.8. It is of spectral type K0 III–IV and lies 75 ly away. β, at magnitude 4.2, is the second-brightest star in Leo Minor. Its spectral type is G8 III–IV, and it lies 100 ly from Earth.

Deep-Sky Objects — Many faint galaxies of 10th magnitude or less.

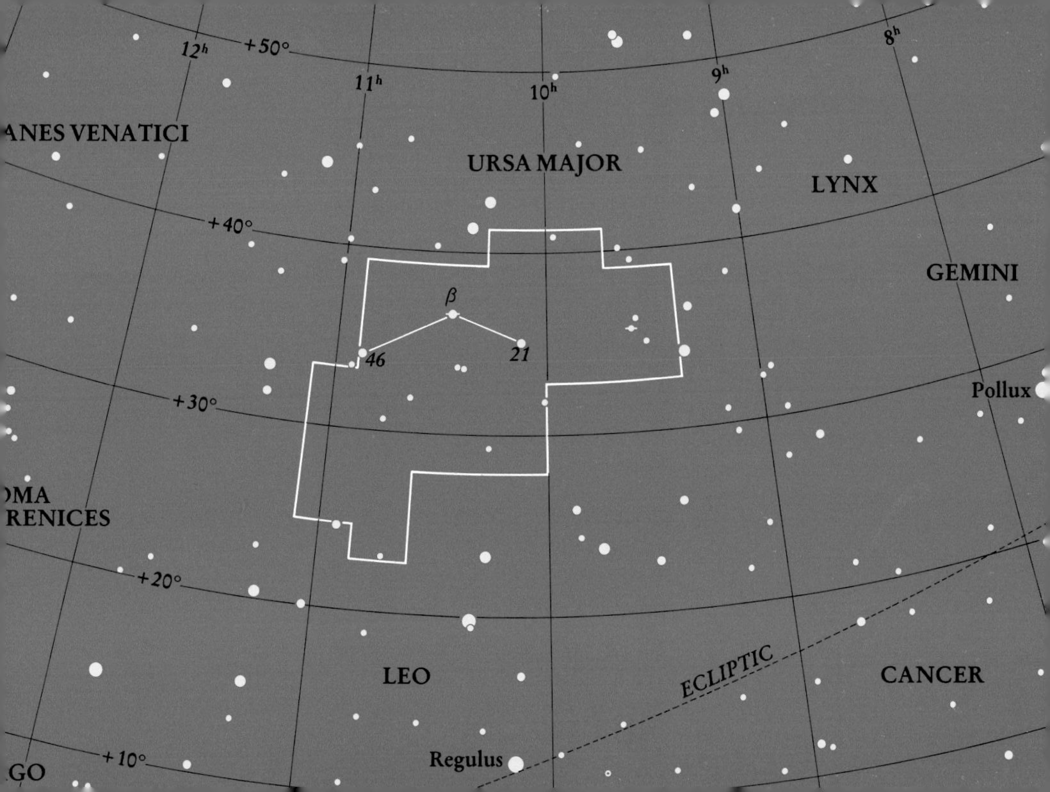
+50°
12h
11h
10h
9h
8h
ANES VENATICI
URSA MAJOR
LYNX
+40°
GEMINI
β
46
21
Pollux
+30°
OMA
RENICES
+20°
LEO
ECLIPTIC
CANCER
+10°
Regulus
GO

Lepus (Lep) The Hare

On Meridian: January 25   Area: 290 square degrees

Despite the faintness of its stars, Lepus is an ancient constellation. It represents a hare or rabbit, a creature associated with the Moon as far back as legends can be traced. Rabbit is said to have been a favorite prey of Orion, so Lepus lies below the Hunter's feet.

Stars R Leporis, the "Crimson Star," is extremely interesting. It is one of the rare so-called carbon stars, very cool and thus very red in appearance. It varies in brightness from magnitude 6.8 to 10.5 and back again over a period of 432 days. When at its brightest it is coppery red. Its color when fainter has been described as wine or blood red. Such red stars as Betelgeuse (α Ori) and Antares (α Sco) seem pale by comparison. It is worth trying to observe R Leporis with binoculars or, preferably, a small telescope. It is located on the western edge of Lepus, about 4° to the northwest of μ Leporis.

Deep-Sky Objects M79 is an 8th-magnitude globular cluster about 5° south and a bit west of β Leporis.

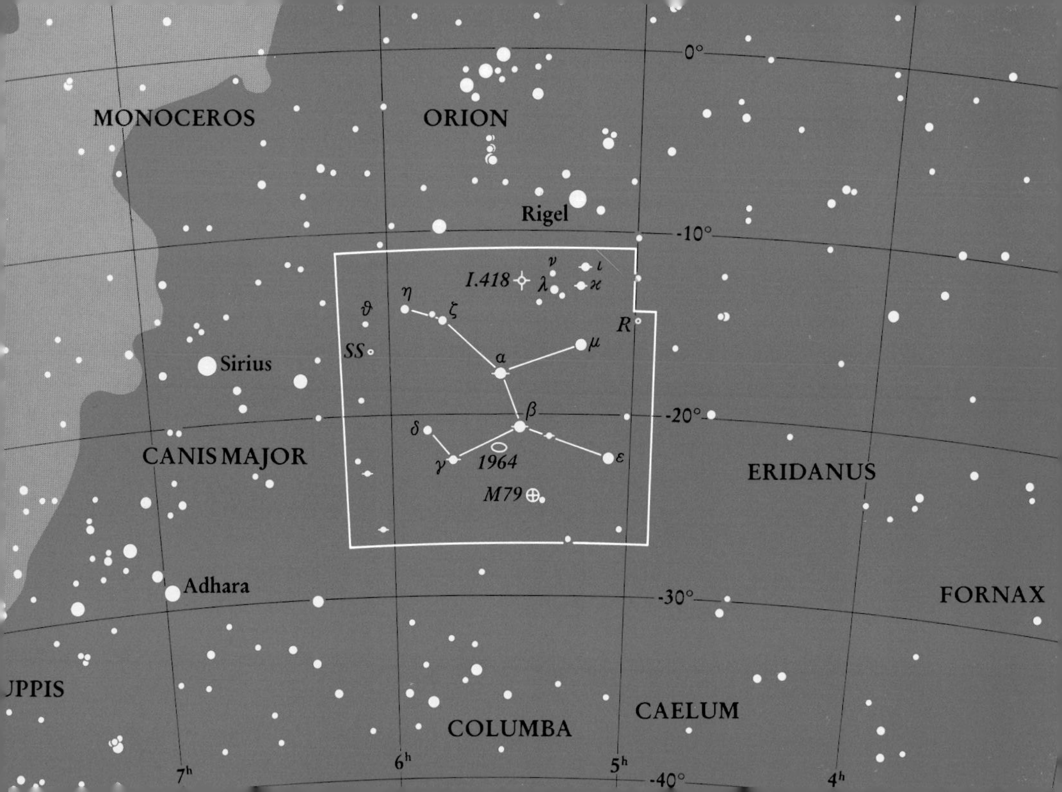
MONOCEROS
ORION
Rigel
0°
-10°
ν
ι
I.418
λ
ϰ
η
ϑ
ζ
R
SS
μ
α
Sirius
β
-20°
δ
CANIS MAJOR
γ
1964
ε
ERIDANUS
M79
Adhara
-30°
FORNAX
JPPIS
CAELUM
COLUMBA
7h
6h
5h
-40°
4h

Libra (Lib) The Scales, the Balance

On Meridian: June 20   Area: 538 square degrees

These stars along the zodiac originally represented the claws of the adjacent constellation Scorpius. The names of Libra's brighter stars recall this past association. Libra probably became a separate constellation during the Roman Empire. At that time the Sun entered Libra on the autumnal equinox (around September 23), when day and night are about equal, and thus the stars came to represent balanced scales.

Stars

α is called Zubenelgenubi, Arabic for "southern claw." It appears to be double but may be an optical binary, two widely separated, unrelated stars that appear close together from our line of sight. The brighter of the two, α^2, is of spectral type A3 III, magnitude 2.8, and lies about 65 ly away. α^1 is of spectral type F5 IV and magnitude 5.2. Its distance is undetermined. β is Zubeneschamali, the "northern claw." Spectral type B8 III; magnitude 2.6; distance 100 ly.

Deep-Sky Objects

None of easy visibility.

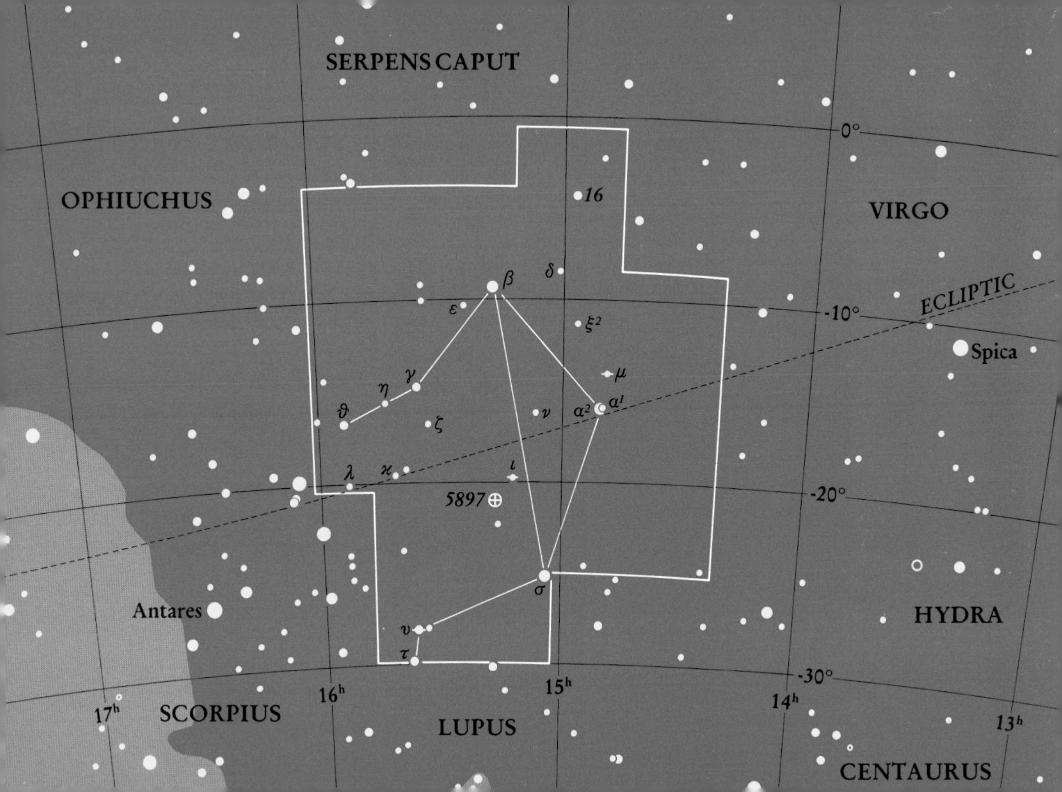

SERPENS CAPUT
OPHIUCHUS
VIRGO
0°
16
β
δ
ε
γ
η
ϑ
ζ
-10°
ξ2
ECLIPTIC
Spica
μ
ν
α2
α1
λ
ϰ
ι
-20°
5897
σ
Antares
HYDRA
υ
τ
-30°
16h
15h
14h
17h
13h
SCORPIUS
LUPUS
CENTAURUS

Lynx (Lyn) The Lynx

On Meridian: March 5   Area: 545 square degrees

This constellation was created about 1687 by Johannes Hevelius to fill in an area southwest of Ursa Major and adjacent to Leo Minor, which he mapped at the same time. He acknowledged the faintness of its stars when he wrote that to see this constellation its observers would have to be "lynx-eyed" and named it after the lynx, a nocturnal wildcat. Lynx spans a fairly large area of sky but contains few deep-sky objects accessible to amateur telescopes. One of its brightest stars is named 10 Ursae Majoris because it was included in the constellation of the Great Bear in John Flamsteed's star catalog, published in 1725.

Stars α is of spectral type K7 III; magnitude 3.1; distance 165 ly. 38 Lyncis, the second-brightest star, is of spectral type A3 V; magnitude 3.8; distance 88 ly. 10 Ursae Majoris, Lynx's third-brightest star, is of spectral type F5 V; magnitude 4.0; distance 46 ly.

Deep-Sky Objects NGC 2683 is a 10th-magnitude spiral galaxy that we see nearly edge-on.

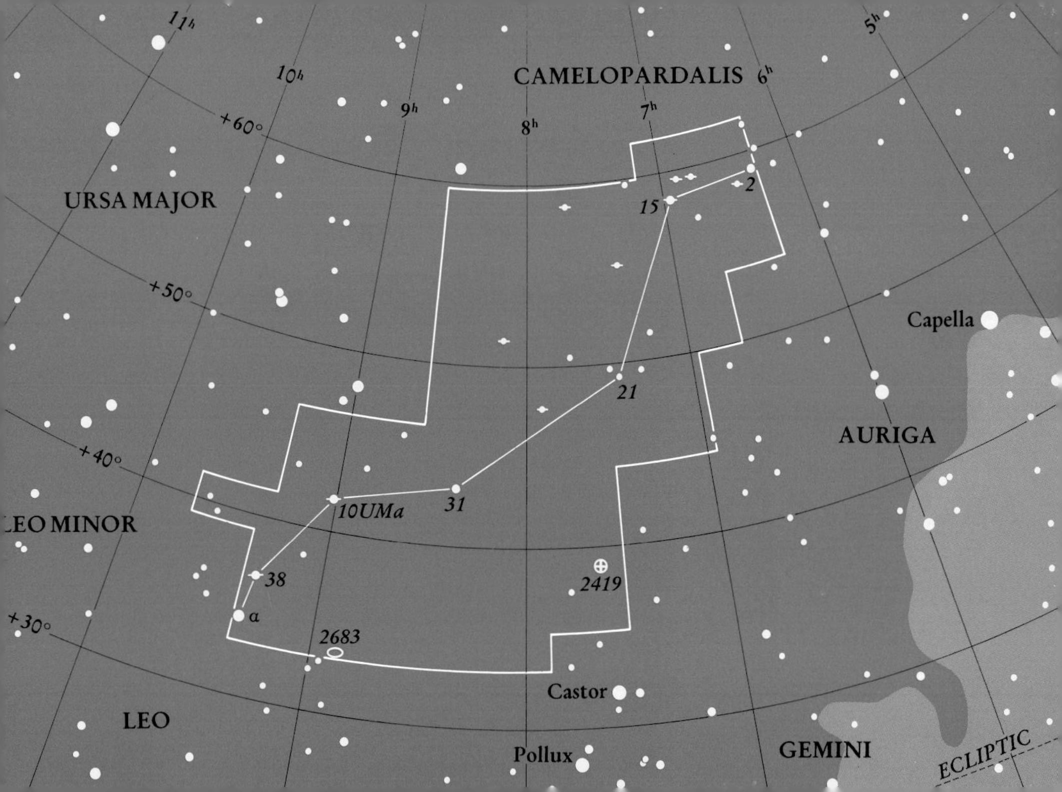

CAMELOPARDALIS
URSA MAJOR
LEO MINOR
LEO
GEMINI
AURIGA
ECLIPTIC
Capella
Castor
Pollux
11h
10h
9h
8h
7h
6h
5h
+60°
+50°
+40°
+30°
15
2
21
31
10UMa
38
α
2419
2683

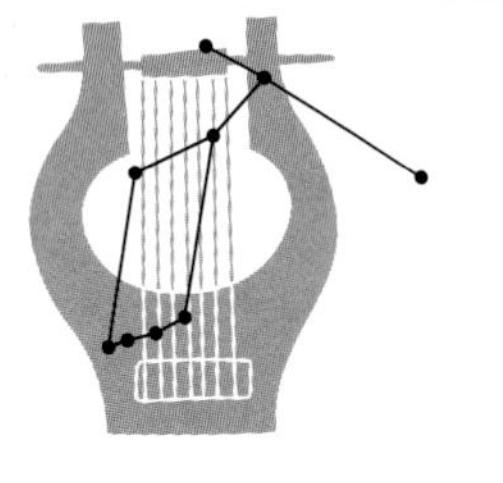

Lyra (Lyr) The Lyre
On Meridian: August 15   Area: 286 square degrees

Among the gems of the sky, Lyra represents the lyre that Orpheus played to persuade the powers of Hades to release his beloved wife, Eurydice. Lyra's brilliant, blue-white star Vega is fifth brightest in the sky.

Stars

α is Vega, "eagle" or "vulture" in Arabic. Spectral type A0 V; magnitude 0.03; distance 26 ly. β, 300 ly from Earth, is the prototype for a class of eclipsing variables. Its two stars, types B7 V and A8, revolve so close to one another that they throw off streams of hot gas. β varies in magnitude from 3.3 to 4.3 over 12.9 days. ϵ appears to the naked eye as a single 4th-magnitude star. Binoculars or a small telescope reveal two stars, of magnitudes 4.7 and 5.1, and a modest-sized telescope reveals that each of these is double.

Deep-Sky Objects

M57, the Ring Nebula, a 9th-magnitude planetary nebula about halfway between β and γ Lyrae, is 2,150 ly away. The "ring" is a shell of gas thrown off by the nebula's central, 15th-magnitude star. M56 is an 8th-magnitude globular cluster.

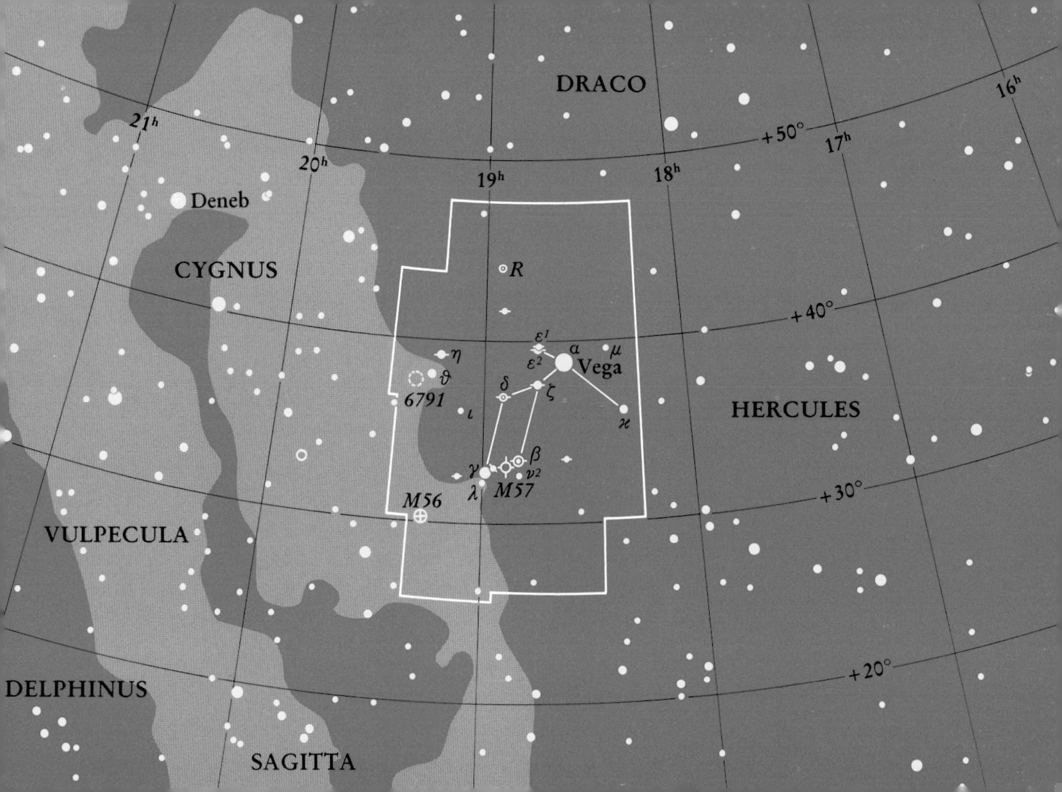
DRACO
16h
17h
18h
19h
20h
21h
+50°
+40°
+30°
+20°
Deneb
CYGNUS
R
ε1
ε2
α
μ
Vega
η
ϑ
6791
δ
ζ
ι
HERCULES
ϰ
β
γ
ν2
λ
M57
M56
VULPECULA
DELPHINUS
SAGITTA

Monoceros (Mon) The Unicorn
On Meridian: February 20   Area: 482 square degrees

The origins of the mythical creature Monoceros—a one-horned animal with the head and forequarters of a horse, the hindquarters of a stag, and the tail of a lion—can be traced back to the Assyrians (2700 B.C. to 600 B.C.). Modern scholars think the idea of the unicorn originated from a mistaken description of the Indian rhinoceros. The constellation is modern, named about 1624 by the German scientist Jakob Bartsch.

Stars

β, a single star of magnitude 3.8 to the naked eye, is a four-star system 715 ly from Earth.

Deep-Sky Objects

M50, a pretty, 6th-magnitude open cluster located near the Canis Major border, can be found about a third of the way along a line from Sirius (α CMa) to Procyon (α CMi). It can be seen with binoculars; a telescope reveals about 100 stars within a region 16′ across. NGC 2244 and NGC 2264 are bright open clusters in northern Monoceros. NGC 2244 is surrounded by the Rosette Nebula, which may be glimpsed in binoculars in excellent, very dark conditions.

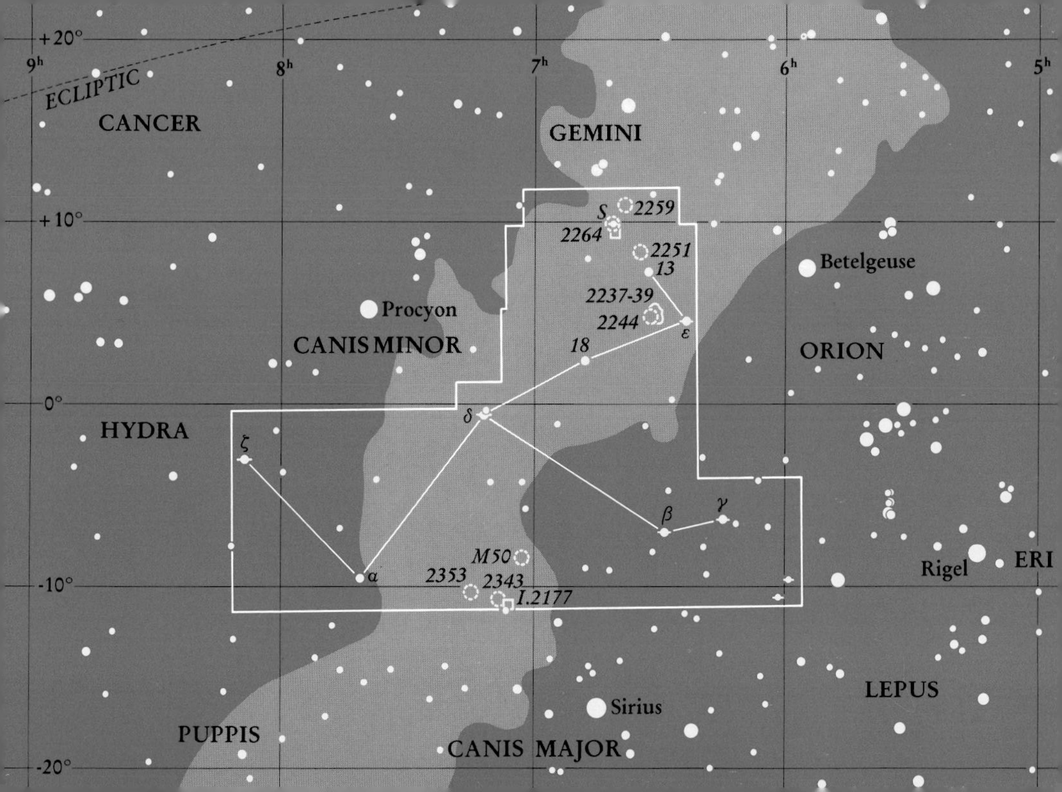

+20°
+10°
0°
-10°
-20°
9h
8h
7h
6h
5h
ECLIPTIC
CANCER
GEMINI
2259
S
2264
2251
13
Betelgeuse
2237-39
Procyon
2244
ε
CANIS MINOR
18
ORION
δ
HYDRA
ζ
β
γ
M50
α
2353
2343
I.2177
Rigel
ERI
LEPUS
Sirius
PUPPIS
CANIS MAJOR

Ophiuchus (Oph) The Serpent Bearer

On Meridian: July 25   Area: 948 square degrees

This large constellation, entwined with the neighboring constellation of Serpens, depicts a man holding a writhing serpent on either side of his body (the constellation Serpens is divided in two parts) and has been known as such for perhaps 4,000 years. Ophiuchus, Greek for "serpent bearer," is usually identified as Asclepius, a legendary physician known as the god of medicine. Asclepius was the son of Apollo and Coronis and was educated by the centaur Chiron.

Stars

α marks the Serpent Bearer's head. Spectral type A5 V; magnitude 2.1; distance 49 ly.

Deep-Sky Objects

IC 4665 is an open cluster 1,400 ly from Earth, found just north of β Ophiuchi. It is of 5th magnitude, with 13 widely spread stars in a region about 1° across. NGC 6633, in the northeastern part of Ophiuchus's border with Serpens Cauda, is a 5th-magnitude open cluster about 20′ across and containing 65 stars. It is about 1,050 ly distant. M9, M10, M12, M14, M19, and M62 are all 7th-magnitude globular clusters.

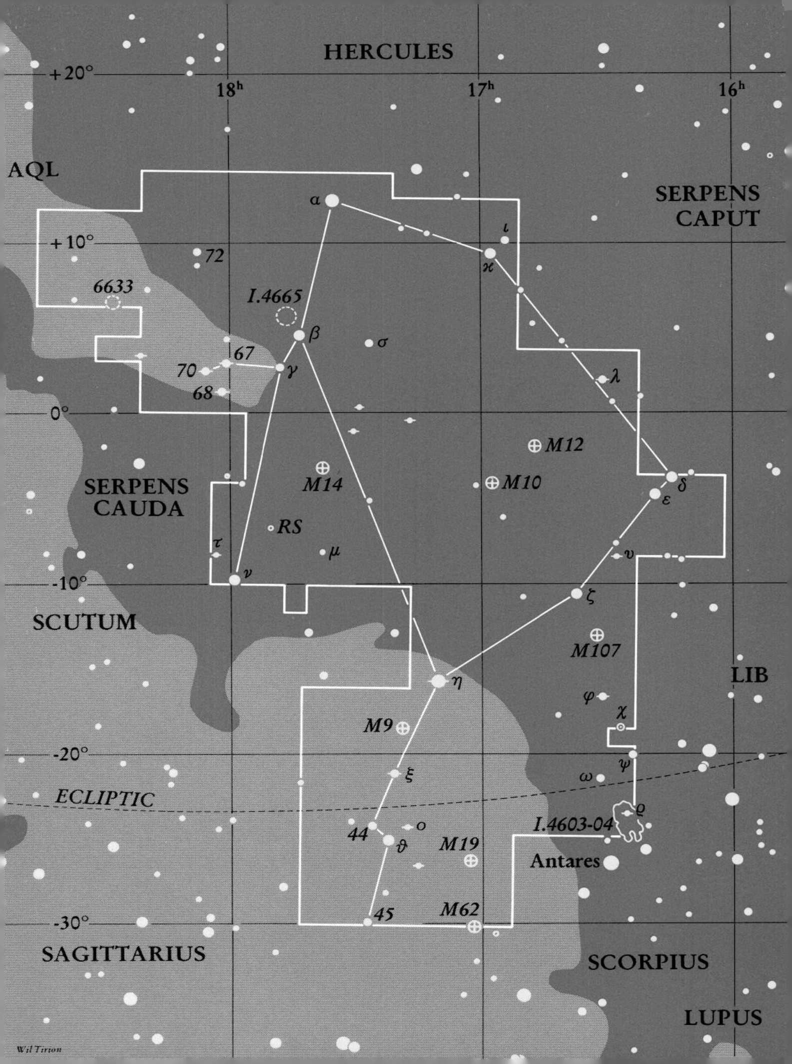
HERCULES
+20°
18h
17h
16h
AQL
SERPENS
CAPUT
α
ι
+10°
72
ϰ
6633
I.4665
β
σ
67
70
γ
λ
68
0°
M12
M14
M10
δ
SERPENS
CAUDA
ε
RS
τ
μ
υ
ν
-10°
ζ
SCUTUM
M107
LIB
η
φ
χ
M9
-20°
ψ
ξ
ω
ECLIPTIC
ϱ
o
I.4603-04
44
M19
ϑ
Antares
45
M62
-30°
SAGITTARIUS
SCORPIUS
LUPUS
Wil Tirion

Orion (Ori) The Hunter

On Meridian: January 25   Area: 594 square degrees

This is perhaps the best-known constellation in the sky. No other constellation looks so like its name or contains so many bright stars. Its stars have been identified as a person—notably the boastful hunter Orion—worldwide for thousands of years.

Stars α, Betelgeuse (from Arabic for "house of the twins," referring to nearby Gemini), 310 ly away, marks Orion's eastern shoulder. It is a type M2 Iab star, larger than the circle inscribed by the orbit of Mars. At magnitude 0.5, it is bright enough for its red color to be apparent to the naked eye. β, Rigel, a brilliant blue-white supergiant of spectral type B8 Ia, 910 ly away, is among the brightest stars in the sky, at magnitude 0.1. ϑ^1, the Trapezium, is a quadruple star system in a cluster of very young stars at the center of the Orion Nebula.

Deep-Sky Objects M42, the Orion Nebula, and its companion, M43, both 1,500 ly away, are below Orion's belt in the sword area. The nebula's greenish clouds and ϑ^1 Orionis are visible in binoculars but are spectacular in a small telescope.

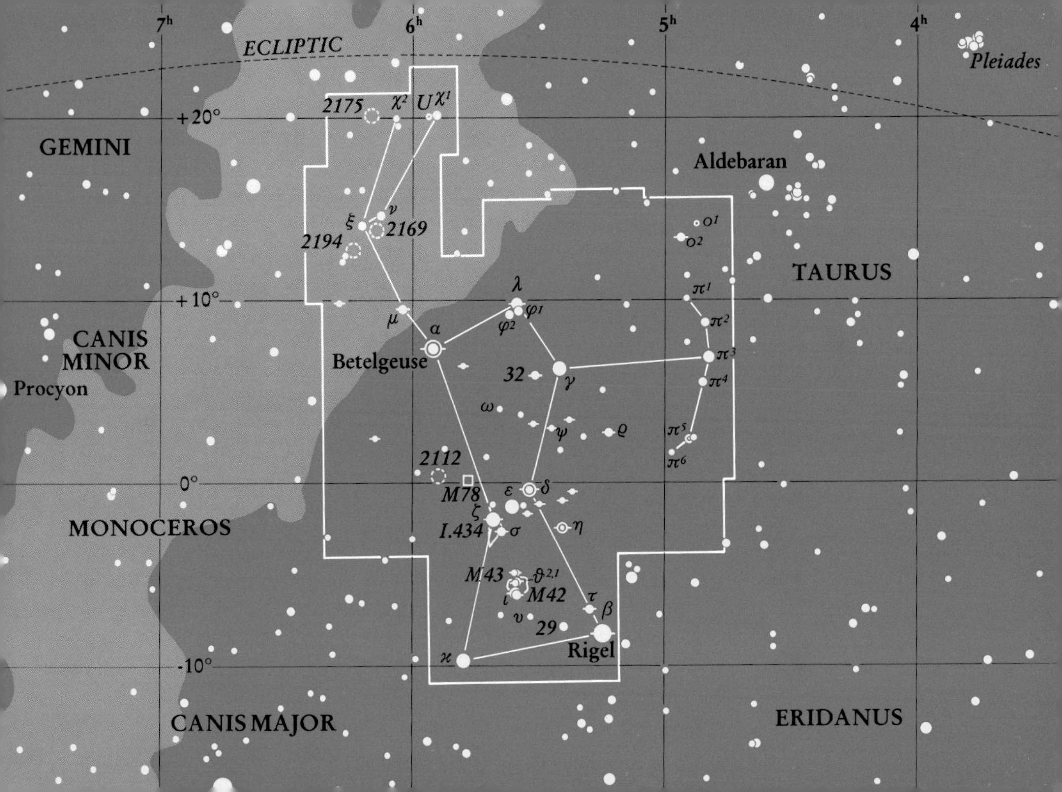
7h
6h
5h
4h
ECLIPTIC
Pleiades
2175
χ2
U χ1
+20°
GEMINI
Aldebaran
ξ
ν
2169
2194
TAURUS
o1
o2
λ
+10°
π1
φ1
π2
μ
α
φ2
CANIS
MINOR
Betelgeuse
π3
32
γ
π4
Procyon
ω
ψ
ρ
π5
π6
2112
0°
M78
ε
δ
ζ
MONOCEROS
I.434
σ
η
M43
ϑ2,1
M42
ι
τ
β
υ
29
Rigel
κ
-10°
CANIS MAJOR
ERIDANUS

Pegasus (Peg) The Winged Horse
On Meridian: October 20   Area: 1,121 square degrees

In Greek mythology Pegasus was a horse born from sea-foam mixed with the blood of the Gorgon Medusa when Perseus severed her head. The constellation's distinctive feature is the Great Square, which is completed by borrowing α Andromedae. Curiously, the horse is upside-down, with its head pointed south.

Stars

α, called Markab, marks the southwestern corner of the Great Square. Spectral type A0 IV; magnitude 2.5; distance 220 ly. β, named Scheat, a slightly variable star in the upper leg of Pegasus, marks the square's northwestern corner. Spectral type M2 II–III; magnitude 2.3–2.7; distance 220 ly. γ, named Algenib, is the southeastern star of the square. Spectral type B2 IV; magnitude 2.8; distance 490 ly.

Deep-Sky Objects

M15, one of the brightest globular clusters in the sky (6th magnitude), is 4° northwest of Enif (ϵ Peg). NGC 7331, the brightest of about a dozen viewable galaxies, is a spiral of 10th magnitude, 9° northwest of β Pegasi, 50 million ly from Earth.

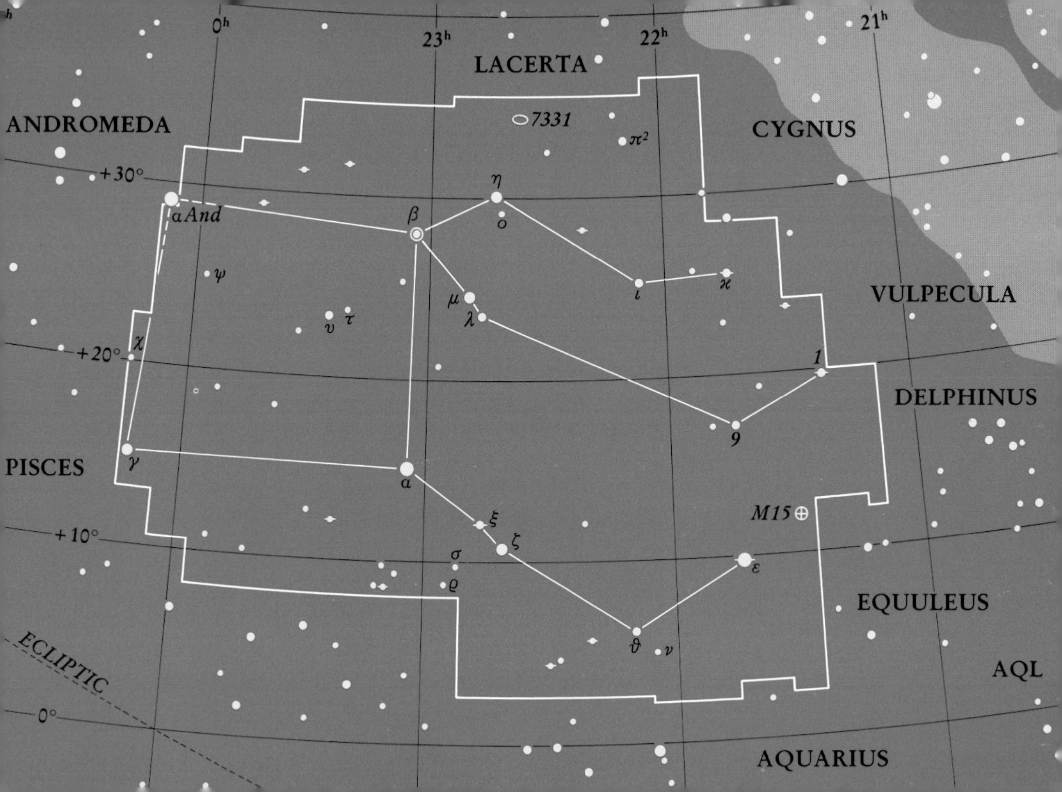

LACERTA
ANDROMEDA
7331
CYGNUS
π^2
+30°
η
α And
β
o
ψ
μ
ι
$\varkappa$
VULPECULA
λ
τ
υ
+20°
χ
1
DELPHINUS
9
PISCES
γ
α
M15
ξ
+10°
ζ
σ
ε
ϱ
EQUULEUS
ϑ
ν
ECLIPTIC
AQL
0°
AQUARIUS
0h
23h
22h
21h

Perseus (Per) The Hero

On Meridian: December 25   Area: 615 square degrees

In Greek mythology Perseus, slayer of Medusa, saved Princess Andromeda, daughter of Cassiopeia and Cepheus, from the sea monster Cetus. The constellation is the radiant (origin) of the Perseid meteor shower of mid-August, which averages 50 "shooting stars" an hour.

Stars

β is Algol, the Demon Star (said to be the eye of Medusa), the prototype of the eclipsing binary class of variables, wherein each member of the star system periodically obscures the other. Algol is composed of a B8 V star in mutual orbit with a G5 IV star, 105 ly away. Usually appearing at a magnitude of 2.1, every 2.87 days Algol "winks," dimming to magnitude 3.3 for 10 hours as the G5 star crosses in front of the B8 star.

Deep-Sky Objects

Located along the Milky Way, Perseus is rich in galactic deep-sky objects. h Persei (NGC 869) and χ Persei (NGC 884) are a pair of 5th-magnitude open clusters known as the Double Cluster, 7,300 ly away, visible as fuzzy stars to the naked eye. M34 is another 5th-magnitude open cluster.

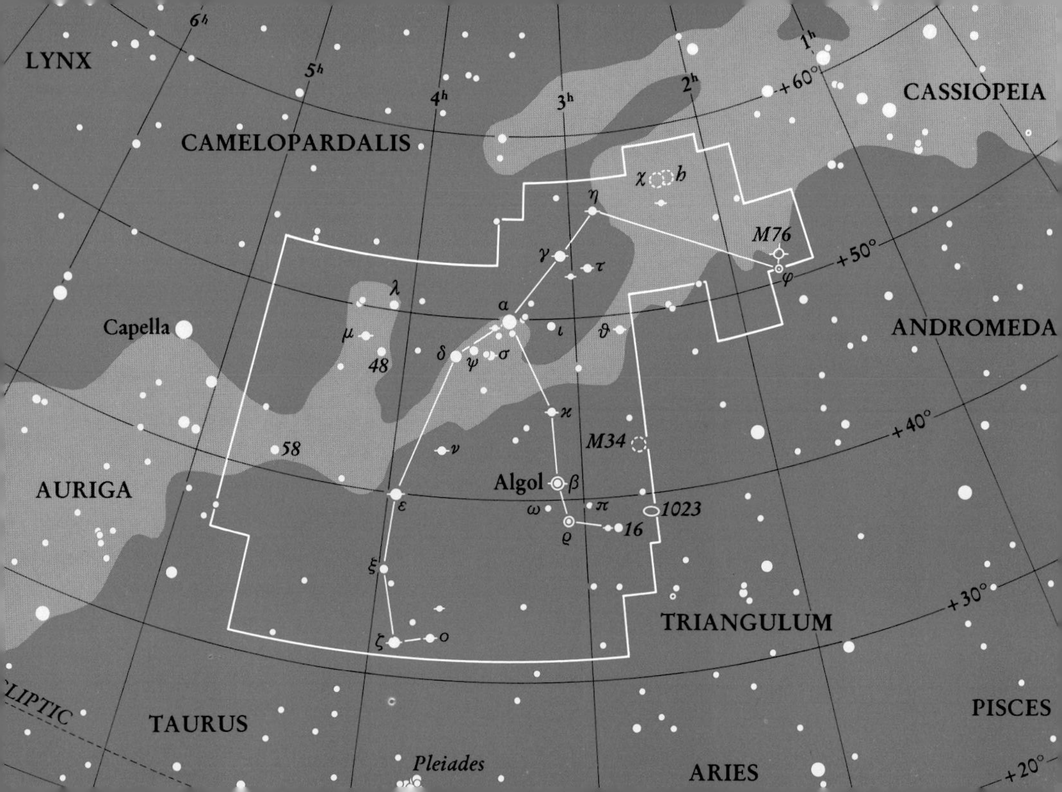

LYNX
CASSIOPEIA
CAMELOPARDALIS
6h
5h
4h
3h
2h
1h
+60°
+50°
+40°
+30°
+20°
χ
h
η
γ
τ
M76
φ
λ
α
Capella
μ
ι
ϑ
ANDROMEDA
δ
ψ
σ
48
ϰ
M34
58
ν
AURIGA
Algol
β
ε
ω
π
1023
ϱ
16
ξ
TRIANGULUM
ζ
o
ECLIPTIC
TAURUS
PISCES
Pleiades
ARIES

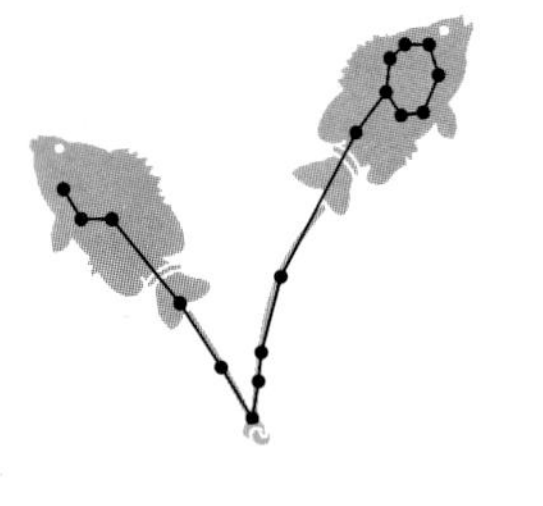

Pisces (Psc) The Fish

On Meridian: November 10   Area: 889 square degrees

Since ancient times the faint stars of the zodiacal constellation of Pisces have represented fish. The Greeks and Romans knew them as Aphrodite and her son Eros, who, as they were fleeing from the monster Typhon, jumped into a stream, turned into fish, and swam away, with their tails tied together. Today the Sun is within Pisces from March 13 to April 19, and so is in the constellation when its path (the ecliptic) crosses the celestial equator around March 21, the vernal equinox, the first day of spring in the Northern Hemisphere.

Stars — α is called Alrisha (also Alrescha), Arabic for "the knot," so called because it ties the fish together. Spectral type A2 V; magnitude 3.9; distance 98 ly.

Deep-Sky Objects — M74 is a nice spiral galaxy of about 9th magnitude located a couple degrees northeast of η Piscium, in the eastern part of the constellation.

TRIANGULUM
ANDROMEDA
+30°
2h
1h
0h
23h
+20°
ARIES
PEGASUS
M74
TV
ECLIPTIC
+10°
UU
TX
0°
Mira
CETUS
AQUARIUS
30

Piscis Austrinus (PsA) The Southern Fish

On Meridian: October 10   Area: 245 square degrees

The constellation Piscis Austrinus has been known since classical Greek and Roman times but probably goes back even further, to an ancient Syrian constellation representing the god Dagon. It has occasionally been shown as two fish, but it is more commonly seen as a single fish, sometimes drinking from a stream of water poured from the jar held by Aquarius, which lies just north of it. Piscis Austrinus has one 1st-magnitude star, Fomalhaut (α PsA), which makes the constellation, just south of Aquarius and Capricornus, fairly easy to find. The rest of its stars are of only 4th magnitude.

Stars

α is called Fomalhaut, a name derived from the Arabic for "fish's mouth." It is of spectral type A3 V and magnitude 1.2. The only 1st-magnitude star in this part of the sky, it is quite easy to spot. At a distance of 22 ly, it is among the nearer stars to our solar system.

Deep-Sky Objects

None of easy visibility.

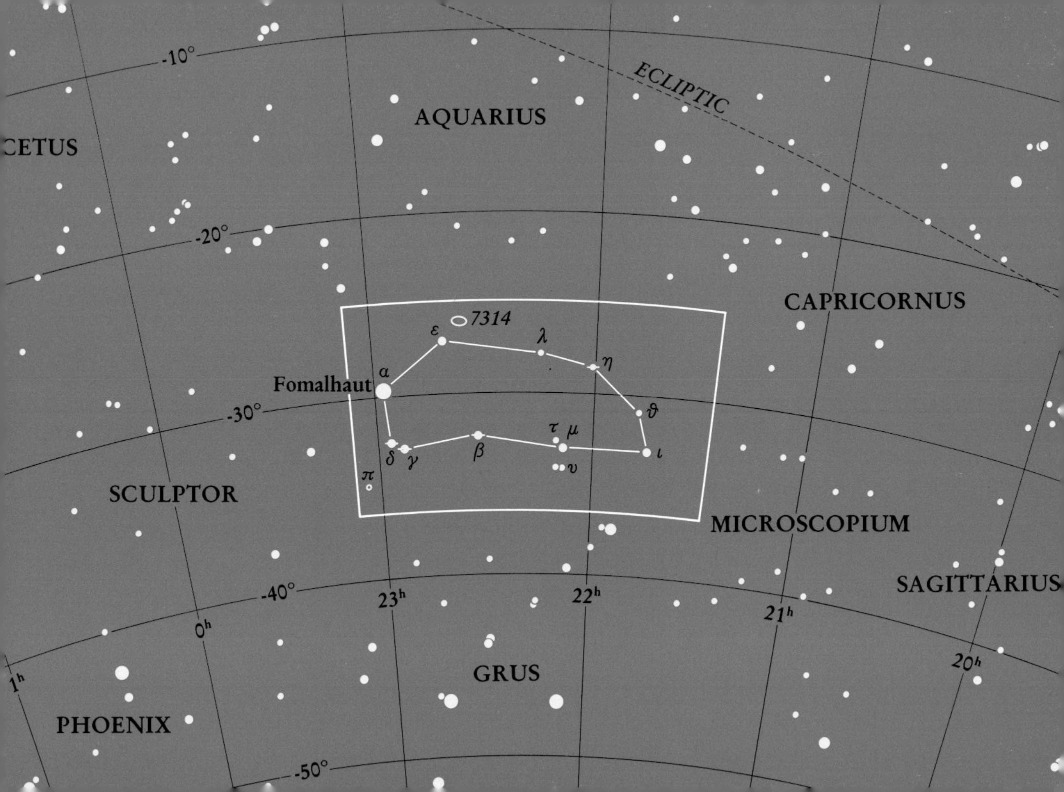
-10°
ECLIPTIC
AQUARIUS
CETUS
-20°
CAPRICORNUS
7314
ε
λ
η
α
Fomalhaut
-30°
ϑ
τ μ
δ γ
β
ι
υ
π
SCULPTOR
MICROSCOPIUM
SAGITTARIUS
-40°
23h
22h
21h
0h
GRUS
20h
1h
PHOENIX
-50°

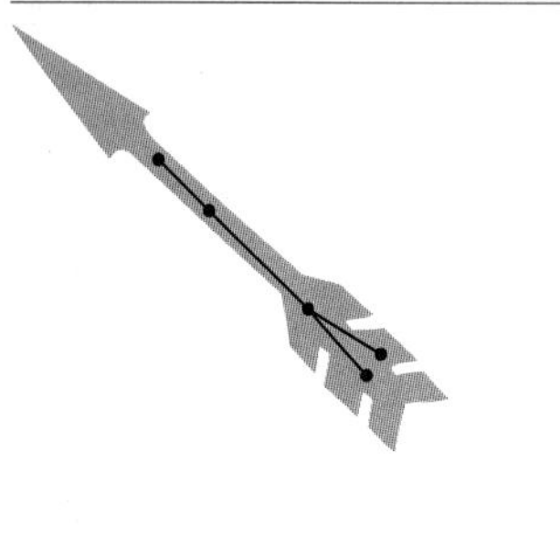

Sagitta (Sge) The Arrow

On Meridian: August 30   *Area: 80 square degrees*

Sagitta, a constellation recognized since classical times, has been identified with just about every famous arrow in mythology. It has been said to be the arrow that killed the eagle of Zeus, the arrow shot by Hercules at the Stymphalian Birds, and the one with which Apollo slew the Cyclops. It has also been said to represent Cupid's arrow. The third-smallest constellation, Sagitta is an inconspicuous area of the sky containing dim stars and few interesting deep-sky objects.

Stars α is sometimes called Sham, from the Arabic word for "arrow." Spectral type G0 II; magnitude 4.4; distance 620 ly.

Deep-Sky Objects M71 is a globular cluster of about 8th magnitude lying between δ and γ Sagittae.

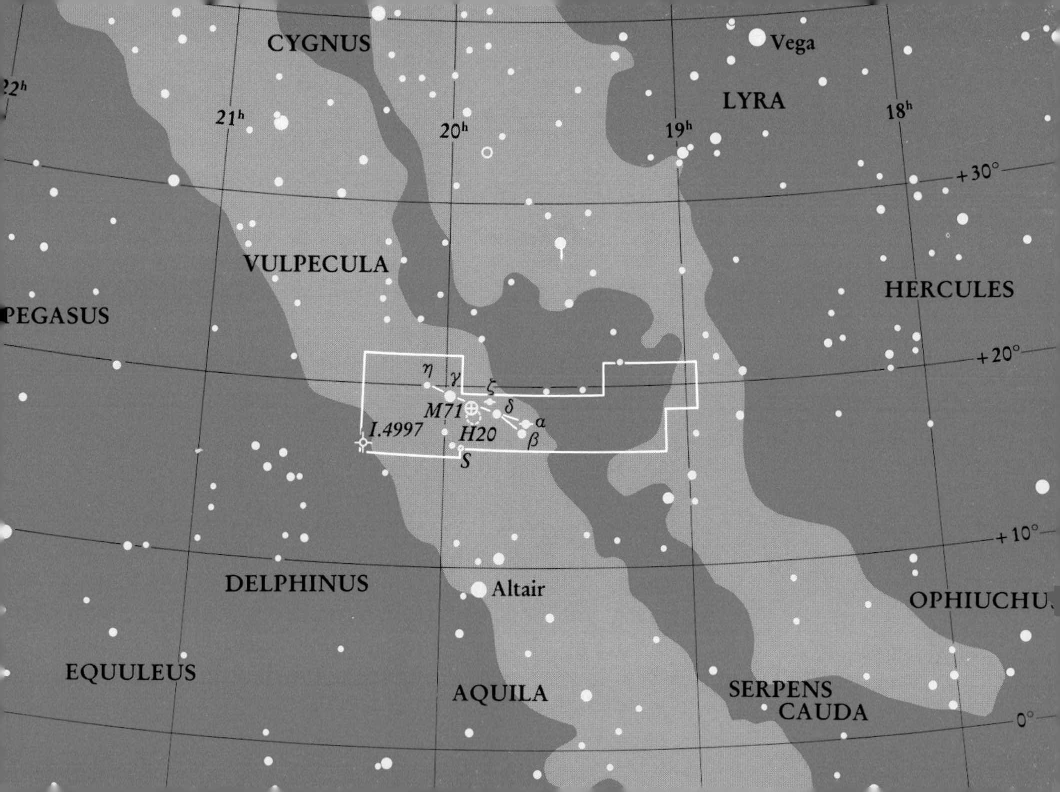
CYGNUS
Vega
LYRA
22h
21h
20h
19h
18h
+30°
VULPECULA
HERCULES
PEGASUS
+20°
η
γ
ζ
δ
α
β
M71
I.4997
H20
S
+10°
DELPHINUS
Altair
OPHIUCHU
EQUULEUS
AQUILA
SERPENS
CAUDA
0°

Sagittarius (Sgr) The Archer
On Meridian: August 20   Area: 867 square degrees

This large zodiacal constellation was probably first associated with a Sumerian arrow-shooting god of war, then as a Greek archer, and later as a satyr or centaur. It is difficult to recognize a centaur, but a modern asterism called the Teapot is easy to find. Sagittarius lies in the direction of our galaxy's center, so the band of the Milky Way is brightest here.

Stars

ϵ is Kaus Australis, a B9 IV star, 85 ly away. At magnitude 1.9, it is the brightest star of Sagittarius.

Deep-Sky Objects

Many deep-sky objects adorn Sagittarius, including 15 Messier objects and the Milky Way's brightest star clouds. M22 is a 5th-magnitude globular cluster about 3° northeast of λ Sagittarii. M23 is a large open cluster of about 120 7th-magnitude stars. M8, the Lagoon Nebula, a large, beautiful cloud surrounding young stars 15,000 ly from Earth, has an overall magnitude of 6.0. Nearby is M20, the Trifid Nebula, containing very hot young stars amid its gas and dust. M17, the Omega Nebula, has a magnitude of 7.0 and lies 10,000 ly from Earth.

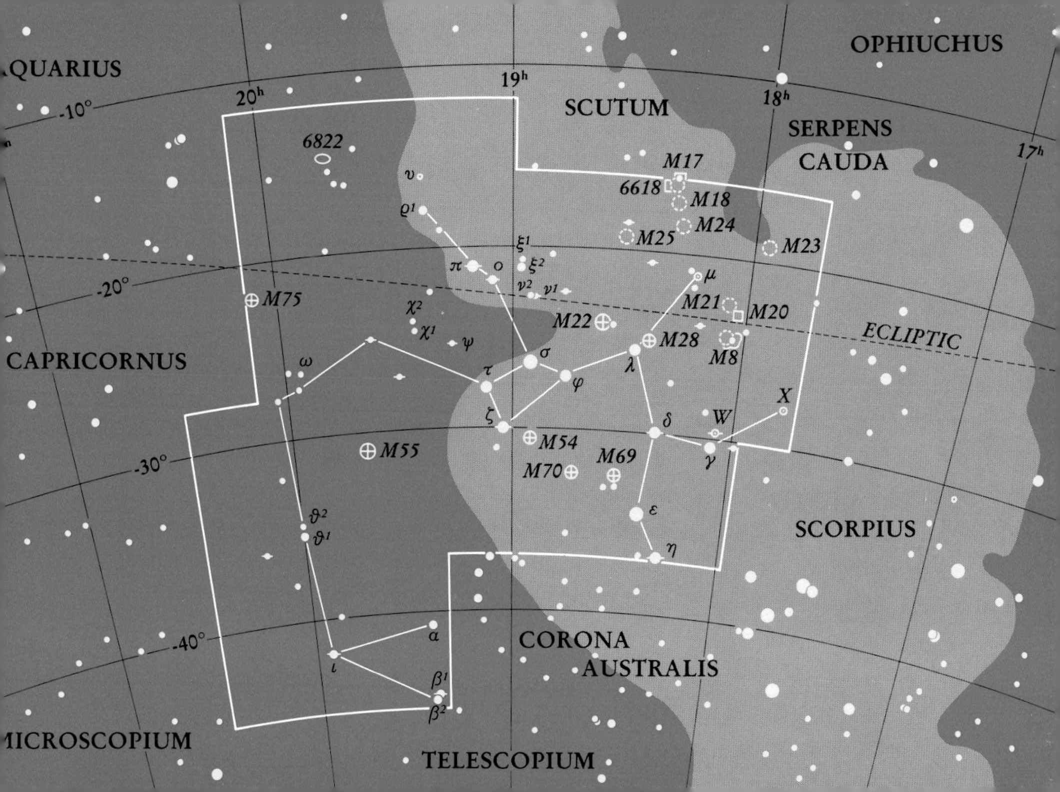

OPHIUCHUS
AQUARIUS
19h
20h
18h
17h
-10°
-20°
-30°
-40°
SCUTUM
SERPENS CAUDA
6822
M17
6618
M18
M24
M25
M23
M75
M22
M21
M20
M28
M8
ECLIPTIC
CAPRICORNUS
X
W
M54
M55
M69
M70
SCORPIUS
CORONA AUSTRALIS
MICROSCOPIUM
TELESCOPIUM

Scorpius (Sco) The Scorpion
On Meridian: July 20   Area: 497 square degrees

This zodiacal constellation is supposed to be the tiny scorpion that killed Orion with its sting. The two constellations are on opposite sides of the sky: Scorpius rises as Orion sets. Scorpius lies along the Milky Way, next to Sagittarius, and so is rich in interesting objects.

Stars

α is Antares, which means "rival of Mars," so named because its noticeably red color rivals that of the red planet. It is a supergiant approximately 600 times the diameter of our Sun, of spectral type M1 Ib and magnitude 1.0. It has a bluish 5th-magnitude companion of spectral type B3. They lie 325 ly away.

Deep-Sky Objects

Scorpius contains more than two dozen star clusters, most of 7th or 8th magnitude, lying on either side of the scorpion's tail. M7 (3rd magnitude) and M6 (4th magnitude) are bright open clusters; M4 is a 6th-magnitude globular cluster.

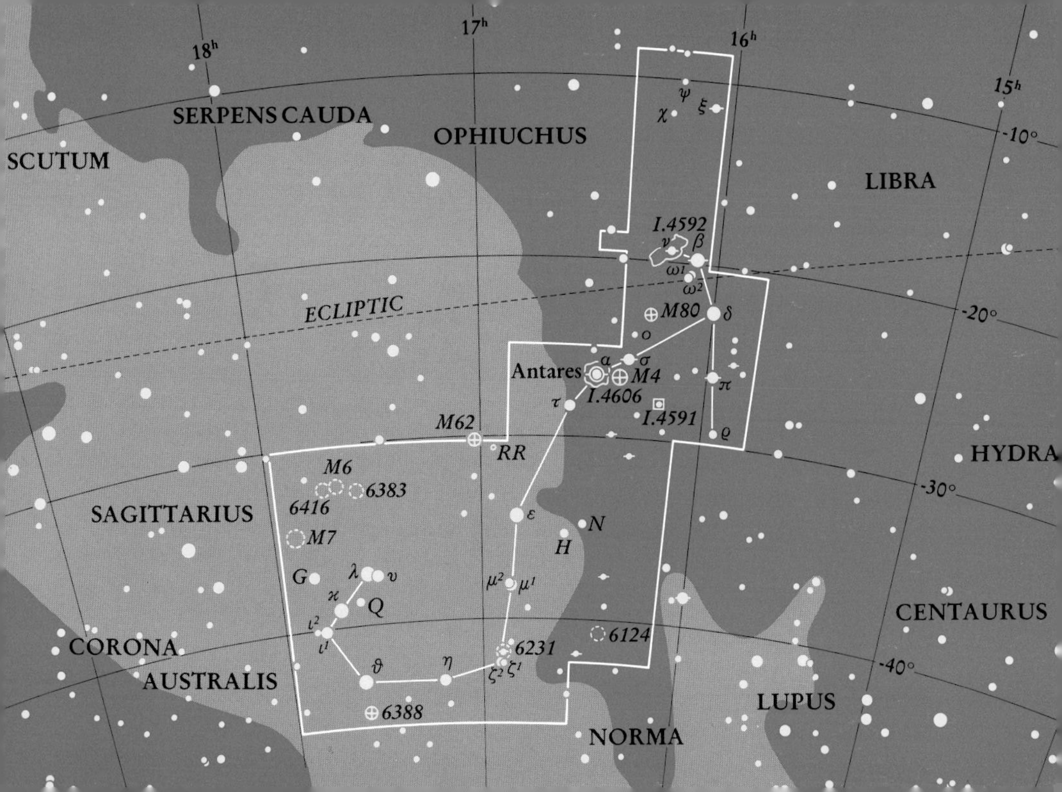

18h
17h
16h
15h
SERPENS CAUDA
OPHIUCHUS
SCUTUM
LIBRA
-10°
-20°
-30°
-40°
ECLIPTIC
I.4592
ν
β
ω1
ω2
M80
δ
ψ
ξ
χ
o
α
σ
Antares
M4
π
τ
I.4606
I.4591
ρ
M62
RR
HYDRA
M6
6383
6416
SAGITTARIUS
M7
ε
N
H
G
λ
υ
κ
Q
μ2
μ1
ι2
ι1
CORONA
AUSTRALIS
6124
6231
ϑ
η
ζ2
ζ1
CENTAURUS
LUPUS
6388
NORMA

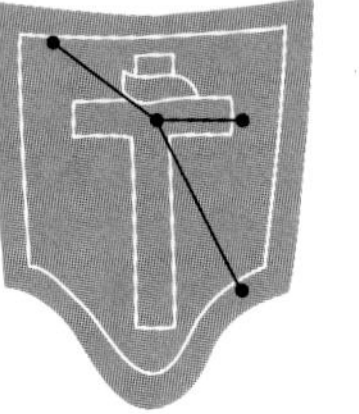

Scutum (Sct) The Shield
On Meridian: August 15   Area: 109 square degrees

This modern constellation was first called Scutum Sobiescianum, "Sobieski's Shield," in honor of 17th-century King John III Sobieski of Poland, and was supposed to represent his coat of arms. Today the name has been shortened to Scutum, the Shield. Although made up of faint stars, Scutum lies close to the center of the Milky Way and contains several star clusters and two Messier objects. This chart also features Serpens Cauda, the Tail of the Snake. It is discussed with Serpens Caput, the Head of the Snake, which follows.

Stars

α: spectral type K3 III; magnitude 3.9; distance 180 ly.

Deep-Sky Objects

M11 is a 6th-magnitude open cluster 5,600 ly away, located about 2° southeast of β Scuti. It has about 200 stars, all between magnitudes 9 and 14, in a region about 10′ across. M26, about 5° south of β Scuti, is a 9th-magnitude open cluster of 20 stars about 5,000 ly from Earth.

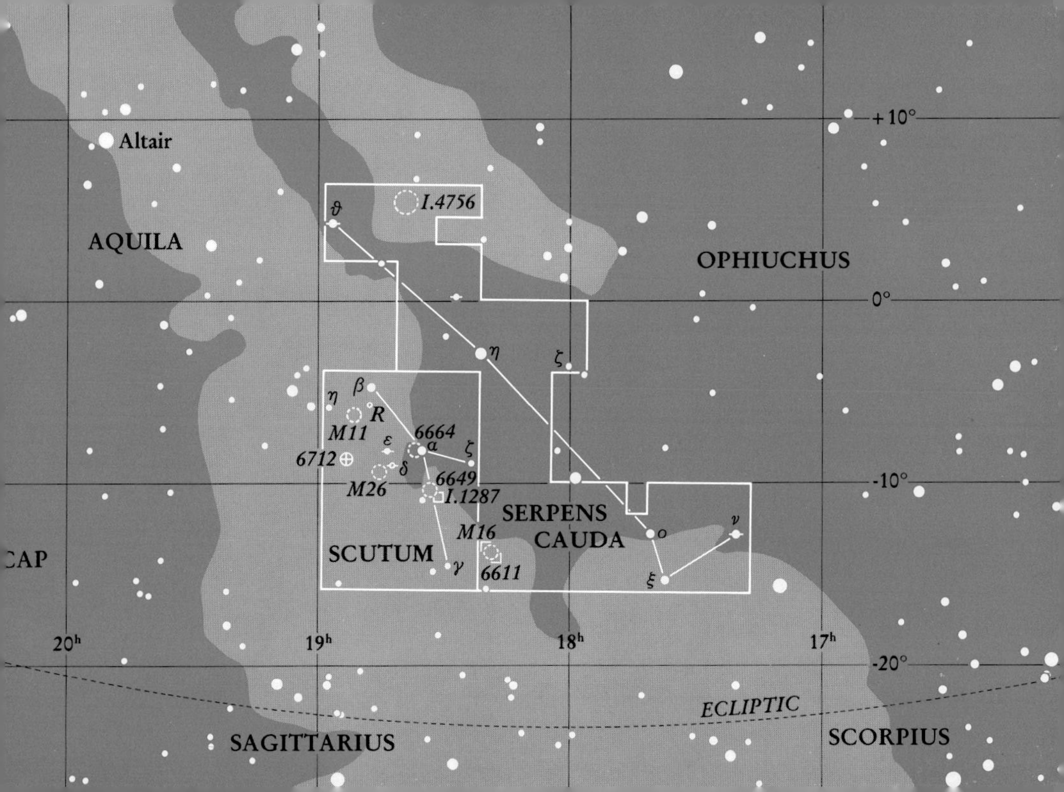
Altair
AQUILA
OPHIUCHUS
I.4756
+10°
0°
-10°
-20°
SERPENS
CAUDA
SCUTUM
CAP
M11
R
6664
6712
M26
6649
I.1287
M16
6611
20h
19h
18h
17h
ECLIPTIC
SAGITTARIUS
SCORPIUS

Serpens Caput/Cauda (Ser) Head/Tail of the Snake
On Meridian: June 30 (Caput), August 5 (Cauda)
Area: 429 (Caput), 208 (Cauda) square degrees

Serpens is the only constellation in two separate parts, its head and tail separated by Ophiuchus, the Serpent Bearer. Originally the group made up one very large constellation. The star names for the snake were assigned as though it were one contiguous area. This sky chart shows Serpens Caput, the head, only; see the Scutum chart for Serpens Cauda, the tail.

Stars — α is in Serpens Caput. Spectral type K2 III; magnitude 2.7; distance 62 ly.

Deep-Sky Objects — M5, a 6th-magnitude globular cluster in Serpens Caput about 8° southwest of α Serpentis, is a fine object for viewing. It is 26,000 ly from Earth. IC 4756 is a 5th-magnitude open cluster 1,400 ly away, about 4° northwest of ϑ Serpentis in the northern part of Serpens Cauda. M16, in the southeastern corner of Serpens Cauda near its border with Sagittarius, is a 7th-magnitude open cluster with nebulosity. It has about 55 stars in a region about ½° across, 5,500 ly away.

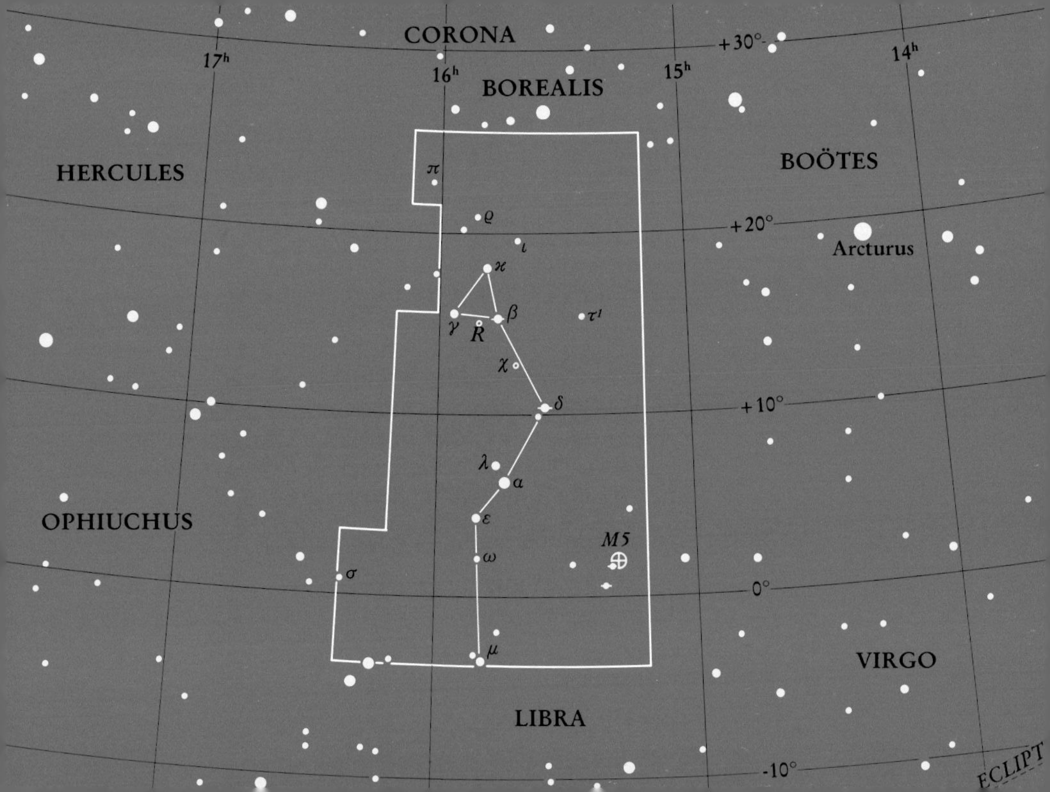
CORONA
BOREALIS
17h
16h
15h
14h
+30°
HERCULES
BOÖTES
+20°
Arcturus
+10°
OPHIUCHUS
M5
0°
VIRGO
LIBRA
-10°
ECLIPT

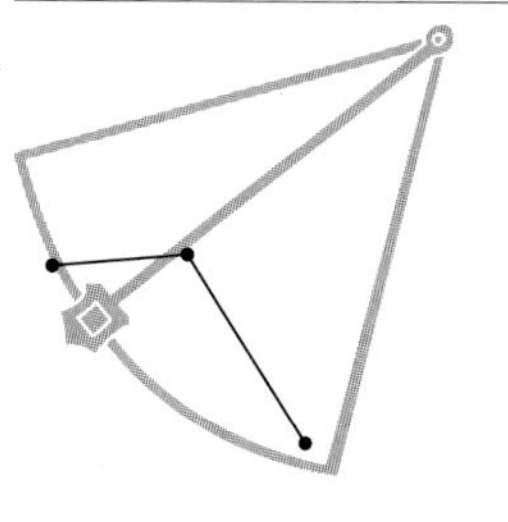

Sextans (Sex) The Sextant

On Meridian: April 5   *Area: 314 square degrees*

This modern constellation honors the astronomical sextant used by the German-Polish astronomer Johannes Hevelius to compile one of the first accurate star maps and detailed charts of the Moon. Hevelius's wife assisted in the observations and published the results—a catalog and atlas of 1,564 stars—in 1690, three years after her husband's death. In the catalog and atlas Hevelius gave shape to several new constellations in areas of the sky that previously had been considered "unformed." Seven of these, including Sextans, continue to be officially recognized today. Sextans is an obscure and relatively uninteresting constellation south of Leo; to find it, look just east of Alphard, Hydra's α star.

Stars — None brighter than 5th magnitude.

Deep-Sky Objects — Sextans contains several galaxies, most fainter than 12th magnitude.

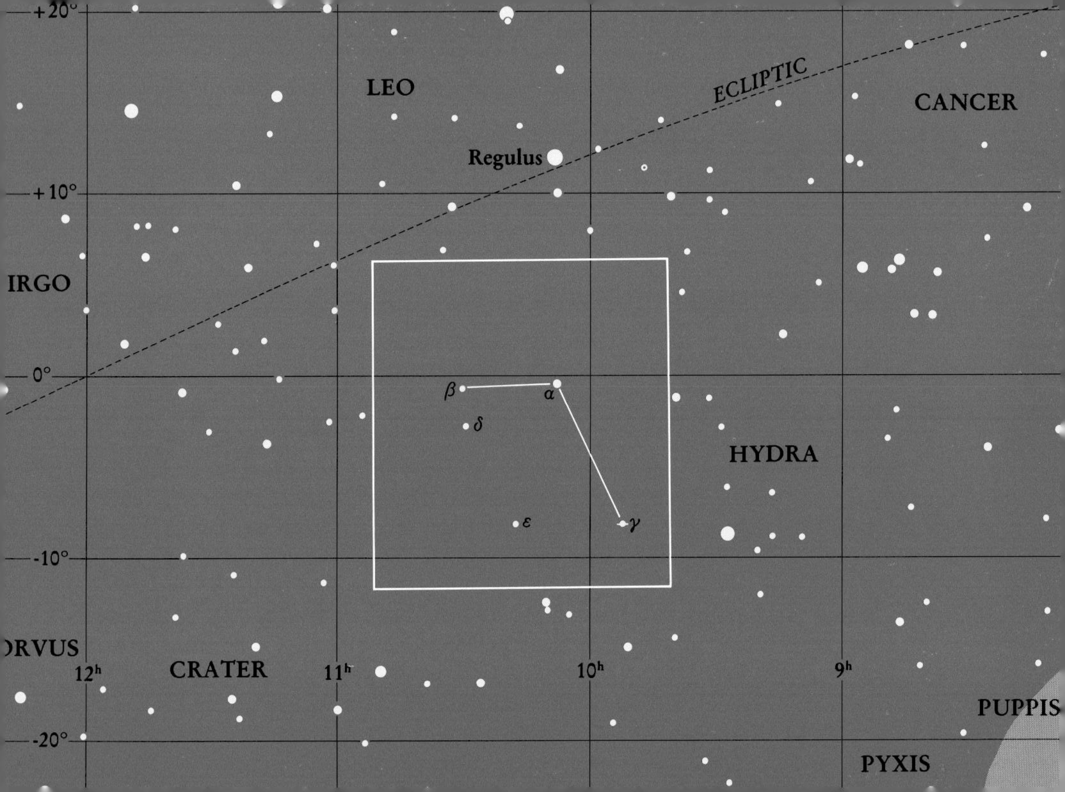
+20°
LEO
ECLIPTIC
CANCER
Regulus
+10°
IRGO
0°
β
α
δ
HYDRA
ε
γ
-10°
ORVUS
CRATER
12h
11h
10h
9h
PUPPIS
-20°
PYXIS

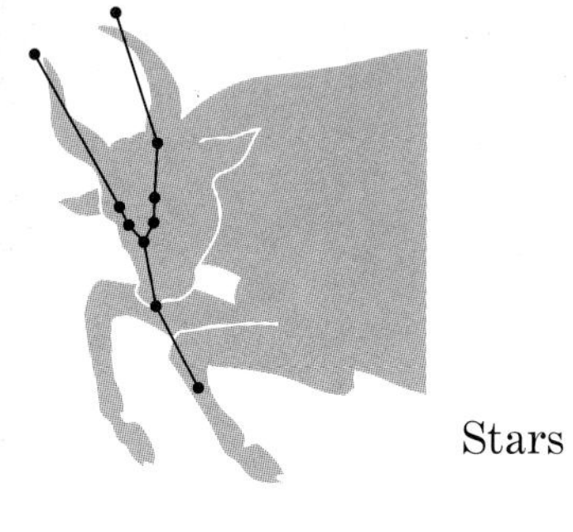

Taurus (Tau) The Bull

On Meridian: January 15   Area: 797 square degrees

Taurus, one of the first constellations recognized, represents a bull, a prominent symbol of strength and fertility in the mythology of many civilizations. Home of the Pleiades and Hyades star clusters and a member of the zodiac, Taurus is of great interest to stargazers.

Stars

α, Aldebaran, Arabic for "the follower" (of the Pleiades), marks the eye of the bull and appears to belong to the Hyades cluster, but is unrelated. A type K5 III star of magnitude 0.9, only 60 ly from Earth, Aldebaran is bright enough to show its orange-red color.

Deep-Sky Objects

The Hyades, a group of about 200 stars in an area 6° across and only 140 ly from Earth, is easily visible to the naked eye as the letter V that forms the face of the bull. The Pleiades (M45), the Seven Sisters of Greek mythology, is a rather young open cluster of hundreds of stars in an area covering 2°, about 415 ly away. In one of the bull's horns lies the most famous supernova remnant known, the Crab Nebula (M1), an 8th-magnitude veil of gas surrounding a superdense, rotating pulsar.

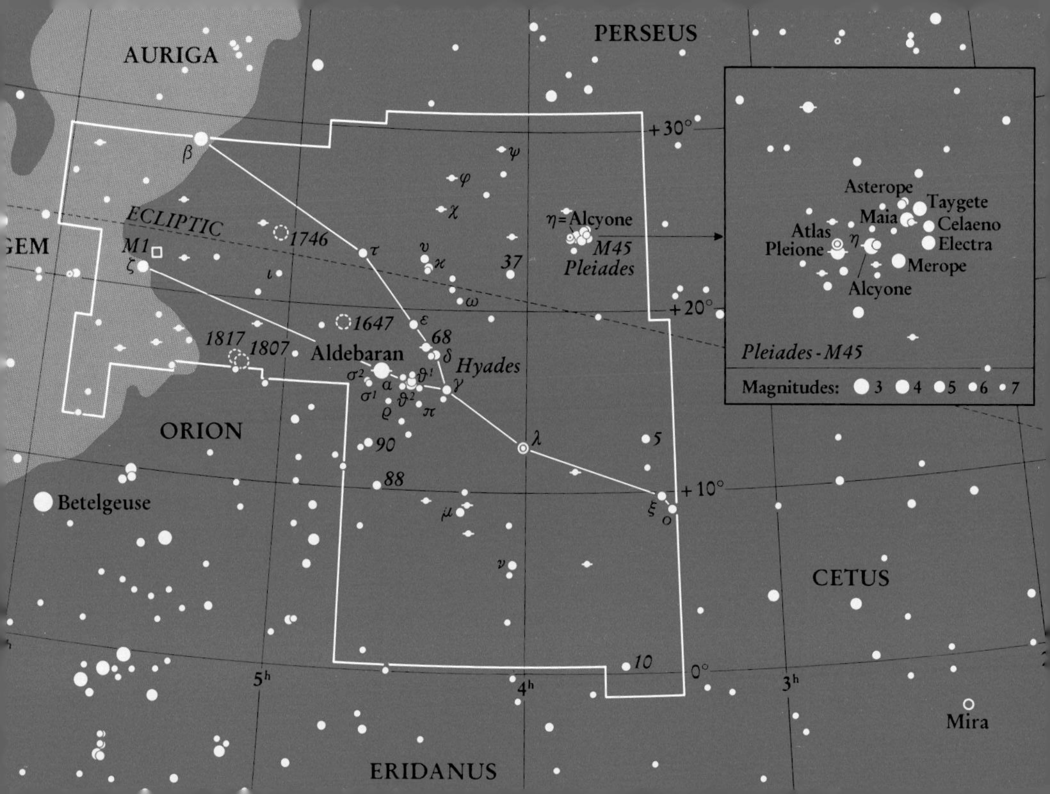

AURIGA
PERSEUS
GEM
ECLIPTIC
M1
ζ
β
τ
ι
1746
ψ
φ
χ
υ
κ
37
η=Alcyone
M45
Pleiades
ω
ε
1647
68
δ
Hyades
1817
1807
Aldebaran
σ2
α
ϑ1
ϑ2
σ1
γ
ρ
π
ORION
90
λ
5
88
+10°
+20°
+30°
ξ
ο
Betelgeuse
μ
ν
CETUS
10
0°
5h
4h
3h
Mira
ERIDANUS
Asterope
Taygete
Maia
Celaeno
Atlas
Electra
Pleione
η
Merope
Alcyone
Pleiades - M45
Magnitudes: 3 4 5 6 7

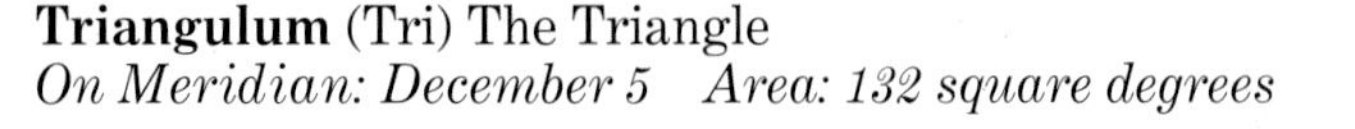

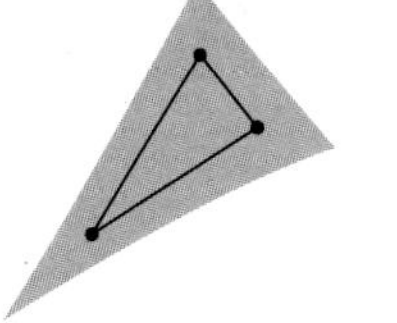

Triangulum (Tri) The Triangle
On Meridian: December 5   Area: 132 square degrees

Just why the ancients should have chosen three faint stars and named them is not known, but this small, dim group has been recognized since classical times. To some it may have represented the delta of the Nile River, to others the triangle-shaped island of Sicily.

Stars — α: spectral type F6 IV; magnitude 3.4; distance 53 ly.

Deep-Sky Objects — M33, a member of our Local Group of galaxies and, at a distance of 2.2 million ly, probably the second-closest spiral to the Milky Way, is among the brightest galaxies accessible to small telescopes. Located near Triangulum's western border with Pisces, it is of about 6th magnitude and $2° \times 1.5°$ in size. Because it is so big, its light is rather diffuse, causing low apparent brightness; a clear, dark sky is usually needed for good viewing. Some sharp-eyed observers report that M33 can be glimpsed with the naked eye on perfectly suited nights.

Capella
5h
AURIGA
4h
3h
2h
1h
0h
CASSIOPEIA
+50°
Algol
+40°
PERSEUS
ANDROMEDA
β
δ
γ
R
ε
M33
α
+30°
Pleiades
PEGASUS
ARIES
+20°
TAURUS
ECLIPTIC
PISCES
+10°

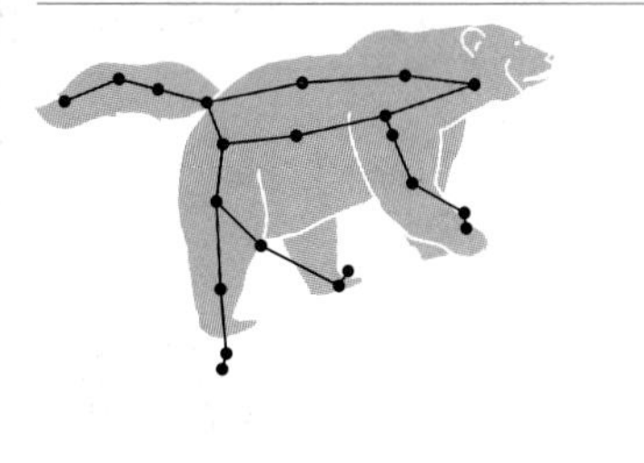

Ursa Major (UMa) The Great Bear

On Meridian: April 20   Area: 1,280 square degrees

Next to Orion, Ursa Major is probably the best known and most storied constellation. It represented a bear to ancient Greeks and Native Americans. The seven bright stars that form the Big Dipper are also called the Plough and the Wain (Wagon).

Stars

α is Dubhe, Arabic for "bear." Spectral type K0 III; magnitude 1.8; distance 75 ly. Dubhe and Merak (β) are called the Pointers, pointing the way to the North Star (α UMi). ζ (Mizar, the middle star in the dipper's handle) and 80 UMa (Alcor) appear to be a double star but are unrelated. Dim Alcor was used as a test of eyesight by Arabs and Native Americans. Mizar: spectral type A2 V; magnitude 2.3; distance 59 ly. Alcor: spectral type A5 V; magnitude 4.0; distance 82 ly.

Deep-Sky Objects

Ursa Major contains many galaxies. Lying 12 million ly away are M81, a 7th-magnitude spiral galaxy (one of the brightest in the sky) and M82, an 8th-magnitude peculiar galaxy apparently undergoing a titanic explosion at its center. M101 is a lovely 8th-magnitude spiral galaxy.

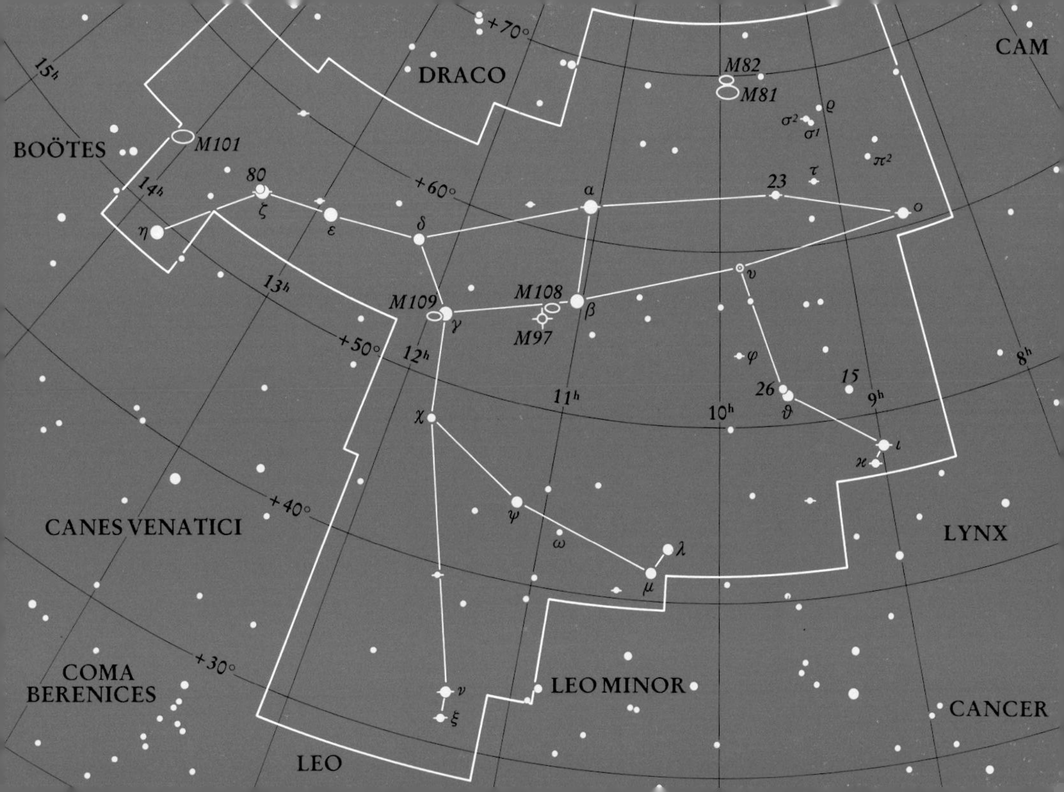

DRACO
CAM
M82
M81
BOÖTES
M101
80
+70°
+60°
14h
15h
23
α
ο
δ
ε
ζ
η
13h
M109
M108
M97
β
γ
υ
+50°
12h
11h
10h
9h
8h
26
ϑ
15
φ
χ
ι
ϰ
+40°
CANES VENATICI
ψ
ω
λ
μ
LYNX
+30°
COMA
BERENICES
LEO MINOR
ν
ξ
CANCER
LEO

Ursa Minor (UMi) The Little Bear

On Meridian: June 25   Area: 256 square degrees

The Little Bear, or Little Dipper, as it is commonly called, contains Polaris, the North Star. Polaris was not always the North Star, however, due to the effects of precession, the cyclical wobbling of the Earth's axis. Four thousand years ago Thuban, the α star in Draco, marked the north celestial pole. Ursa Minor was not recognized as a constellation until about 600 B.C., when it was described by the Greek astronomer Thales.

Stars α is Polaris, the polestar or the North Star. Today it lies slightly less than 1° from the true north celestial pole; it will be closest to the pole in about the year 2105. Many people mistakenly think Polaris is the brightest star in the sky, but, at magnitude 2.0, it is in fact only 49th brightest. It is a slightly variable star (Cepheid type) of spectral type F8 Ib, and lies about 820 ly from Earth. It has an 8th-magnitude, type F3 V companion, visible in a 3″ telescope.

Deep-Sky Objects None notable.

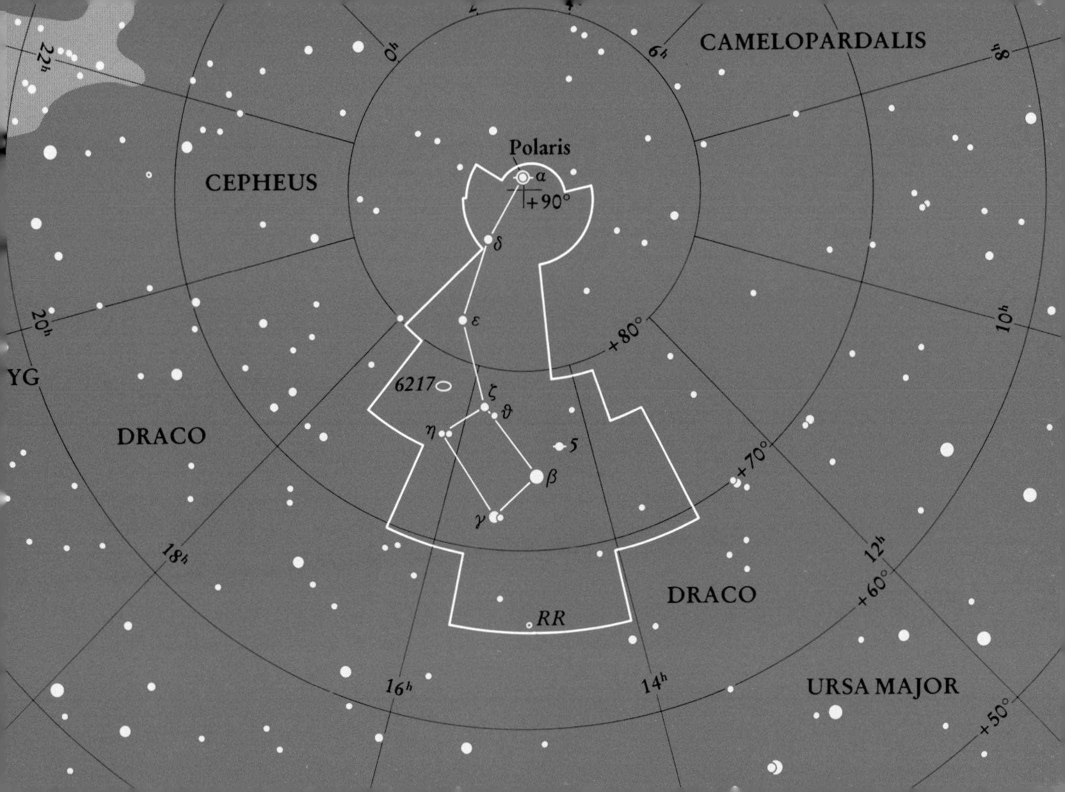

CAMELOPARDALIS
CEPHEUS
Polaris
α
+90°
δ
ε
+80°
6217
ζ
ϑ
η
5
β
γ
DRACO
DRACO
YG
+70°
RR
+60°
URSA MAJOR
+50°
22h
0h
6h
8h
10h
12h
14h
16h
18h
20h

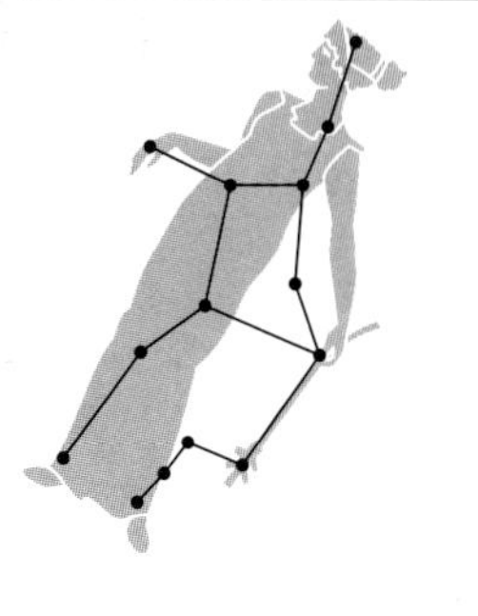

Virgo (Vir) The Maiden
On Meridian: May 25   *Area: 1,294 square degrees*

Virgo, the only female figure in the zodiac, is one of the oldest constellations and has assumed the identity of numerous important female deities of the world, including Ishtar, Isis, Demeter, Athena, and Artemis. Virgo is a huge constellation, second largest in the sky. The Sun spends more time in Virgo than in any other constellation of the zodiac. Virgo's only bright star, Spica, is quite noticeable, seeming to stand alone.

Stars

α is Spica, Latin for "ear of wheat," which is held in the maiden's hand. Spectral type B1 V; magnitude 1.0; distance 220 ly. It has a very faint companion.

Deep-Sky Objects

Within this constellation is the Virgo cluster of galaxies, the central cluster of the Virgo Supercluster, of which our galaxy's cluster, the Local Group, is a member. The Virgo cluster contains perhaps 3,000 galaxies 50 million ly from Earth. A moderate-size telescope is required to see most of them. The brightest are M49, an 8th-magnitude elliptical galaxy, and M104, an 8th-magnitude spiral galaxy.

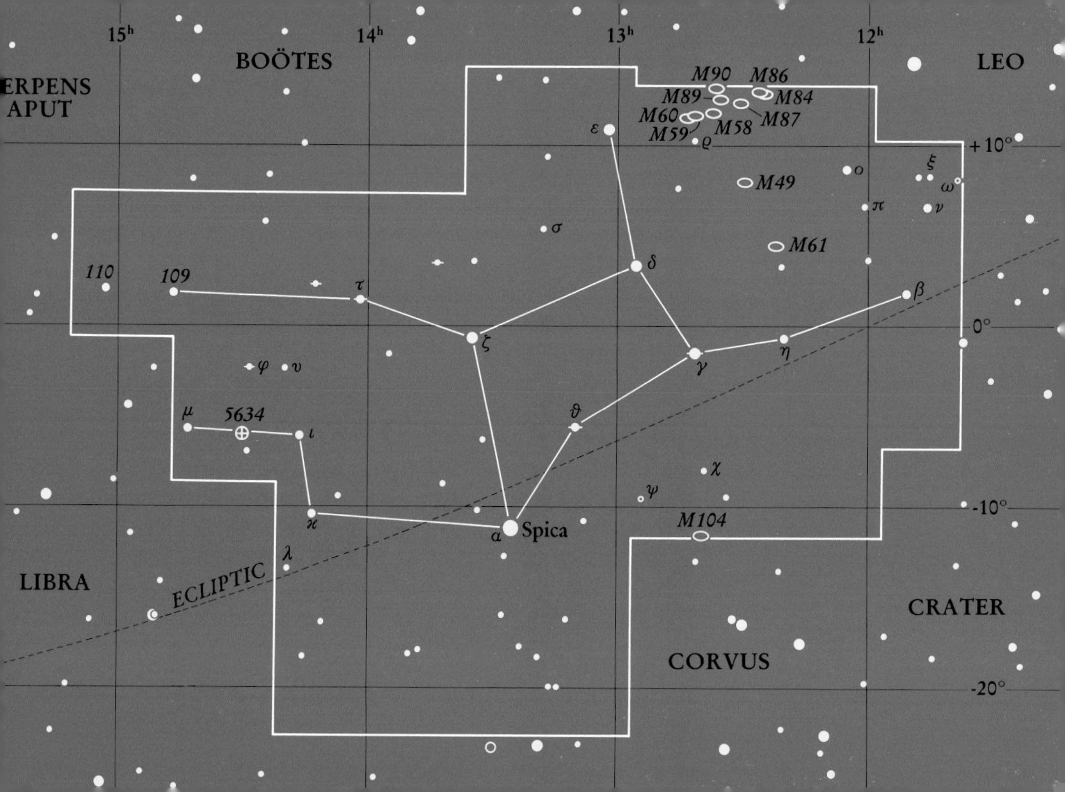

15h
14h
13h
12h
BOÖTES
LEO
ERPENS
APUT
M90
M86
M89
M84
M60
M87
M58
M59
ε
ϱ
+10°
ξ
o
M49
ω
π
ν
σ
M61
δ
110
109
τ
β
0°
ζ
η
γ
φ
υ
μ
5634
ϑ
ι
χ
ψ
-10°
ϰ
M104
α
Spica
λ
LIBRA
ECLIPTIC
CRATER
CORVUS
-20°

Vulpecula (Vul) The Fox

On Meridian: September 10   *Area: 278 square degrees*

Lying just south of Cygnus, Vulpecula is a modern constellation, formed in 1690 by the German-Polish astronomer Johannes Hevelius. Hevelius named it Vulpecula cum Anser, the Fox with the Goose, but the goose has been cooked, and today we are left with only the fox. The constellation contains no bright stars and only one notable deep-sky object, the well-known Dumbbell Nebula.

Stars α is the only Greek-letter star of the constellation. Spectral type M0; magnitude 4.4; distance 85 ly.

Deep-Sky Objects M27, the Dumbbell Nebula, is an 8th-magnitude planetary nebula, so named because the sphere of gas surrounding the central star is brighter on two sides and so appears to have the shape of a dumbbell. It is about 8′ in size, 980 ly from Earth, and can be found about 7° southeast of α Vulpeculae.

LACERTA
Deneb
3h
22h
+40°
21h
20h
19h
Vega
18h
LYRA
CYGNUS
+30°
T
6940
HERCULES
13
α
M27
1
Cr399
+20°
PEGASUS
SAGITTA
+10°
OPH
DELPHINUS
Altair
SERPENS
CAUDA
QUARIUS
EQUULEUS
AQUILA

Constellations of the Southern Skies

The following constellations occur south of −30° declination and so are visible mainly from southern latitudes. Parts of some of these constellations are visible from southern areas of the United States. The chart lists each southern constellation's official Latin name, its common name or identification, and its equatorial coordinates: right ascension (R.A.) and declination (Dec.). See the appendix "Systems of Measurement" for an explanation of celestial coordinates.

Official name	Common name	R.A.	Dec.
Antlia	The Air Pump	10^h	−35°
Apus	The Bird of Paradise	16^h	−75°
Ara	The Altar	17^h	−55°
Caelum	The Chisel	5^h	−40°
Carina	The Keel	9^h	−60°
Centaurus	The Centaur	13^h	−50°
Chamaeleon	The Chameleon	11^h	−80°
Circinus	The Compass	15^h	−60°
Columba	The Dove	6^h	−35°
Corona australis	The Southern Crown	19^h	−40°
Crux	The Southern Cross	12^h	−60°
Dorado	The Goldfish	5^h	−60°
Grus	The Crane	22^h	−45°
Horologium	The Clock	3^h	−45°
Hydrus	The Water Snake	2^h	−75°
Indus	The Indian	21^h	−55°
Lupus	The Wolf	15^h	−45°

Official name	**Common name**	**R.A.**	**Dec.**
MENSA	The Table Mountain	5^h	$-75°$
MICROSCOPIUM	The Microscope	21^h	$-35°$
MUSCA	The Fly	12^h	$-70°$
NORMA	The Level	16^h	$-50°$
OCTANS	The Octant	22^h	$-85°$
PAVO	The Peacock	20^h	$-65°$
PHOENIX	The Phoenix	1^h	$-50°$
PICTOR	The Painter	6^h	$-55°$
PUPPIS	The Stern	8^h	$-40°$
PYXIS	The Compass	9^h	$-30°$
RETICULUM	The Net	4^h	$-60°$
SCULPTOR	The Sculptor	0^h	$-30°$
TELESCOPIUM	The Telescope	19^h	$-50°$
TRIANGULUM AUSTRALE	The Southern Triangle	16^h	$-65°$
TUCANA	The Toucan	0^h	$-65°$
VELA	The Sail	9^h	$-50°$
VOLANS	The Flying Fish	8^h	$-70°$

Systems of Measurement

Many of our modern systems of measurement were derived from methods developed by ancient skywatchers to specify the locations of objects in the sky. Astronomers use systems of measurement to gauge sizes, positions, and distances of celestial objects.

The basic unit of measurement used for celestial distances is the *light-year* (ly), the distance that light travels in a vacuum in one year. Since light travels at a rate of 299,792.458 km per second (186,282 miles per second), in one year it travels 9.46×10^{12} km (5.88×10^{12} miles).

Astronomers have also developed coordinate systems (similar to terrestrial latitude and longitude systems) for plotting objects in the sky and gauging distances between them.

Topocentric Coordinates

One coordinate system you may find useful is the *topocentric, altazimuth,* or *horizon* system. From an observer's point of view this system is perhaps easiest to understand. The coordinates simply measure up and down in the sky, and left-right. The *altitude* of a celestial object is its angular distance above the horizon. An object at 0° altitude is on the horizon; one at 90° altitude is directly overhead at the *zenith.* The *azimuth* tells us the compass direction of an object. Azimuth is measured eastward along the horizon from the north cardinal point (0° azimuth): 90° azimuth is due east, 180° azimuth is due south, and 270° is due west.

Topocentric coordinates of a celestial object are accurate only at a specific date and time, because the positions of celestial objects, as plotted with these Earth-fixed coordinates, are continually changing due to the Earth's motions, mainly rotation. The seasonal sky charts, in the Sky Tours section, for example, show how parts of the sky look at the specific dates and times given. The constellation charts give the date when each constellation culminates at 9:00 P.M. A constellation *culminates* when it crosses the observer's *celestial meridian*—a line that runs from the exact north point on your horizon (0° azimuth) up through the zenith (90° altitude) and down to the south point on the horizon (180° azimuth), dividing the sky into eastern and western halves.

The Celestial Sphere

Other coordinate systems are fixed on the celestial sphere, the "dome" of the heavens. These coordinates do not change with Earth's movements, but the opportunity to observe an object at a designated point in the heavens still depends on the date

Topocentric Coordinates

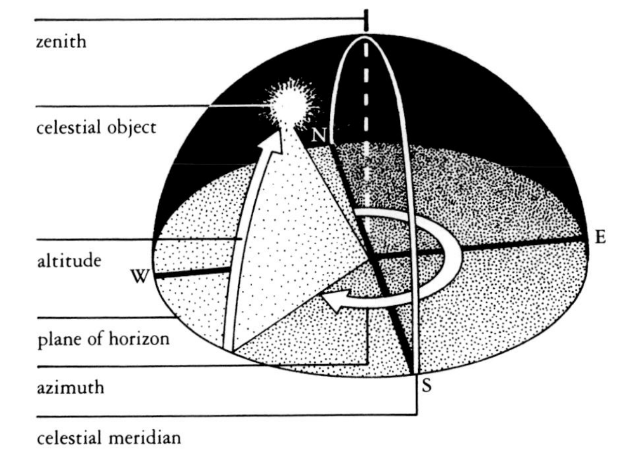

and time of observation. The celestial sphere is an infinitely large imaginary sphere surrounding Earth onto which are "pasted" all the objects seen in the sky. The *celestial equator* is a projection of Earth's equator onto the celestial sphere. Similarly, the *celestial poles* are projections of Earth's north and south geographic poles onto the celestial sphere. The *ecliptic* can be seen as the projection of Earth's orbit onto the celestial sphere or, in terms of the Sun, as the path traced by the Sun around the sky (this path is inclined at 23.4° to the celestial equator because of Earth's 23.4° tilt). The *ecliptic poles* are imaginary lines perpendicular to the ecliptic extending north and south into space.

Equatorial Coordinates

One commonly used set of celestial coordinates, called equatorial coordinates, is based on the celestial equator. The north-south coordinate, the celestial equivalent of latitude on Earth, is called *declination;* it is measured, like latitude, in degrees (°), *minutes of arc* (′), and *seconds of arc* (″), from 0° at the celestial equator to 90° north and 90° south at the celestial poles. The entire celestial sphere is divided into 360°. One degree can be divided into 60 minutes of arc (60′), and one minute of arc can be divided into 60 seconds of arc (″).

The east-west coordinate, the celestial equivalent of longitude on Earth, is called *right ascension;* it is usually measured eastward around the sky in hours (^{h}), minutes (^{m}), and seconds (^{s}) of time but can also be measured in degrees. Since Earth rotates 360° in 24 hours, one hour (1^h) of right ascension, or time, equals 15° of arc; one minute (1^m) of right ascension equals 15 minutes (15′)

The Celestial Sphere and Equatorial Coordinates

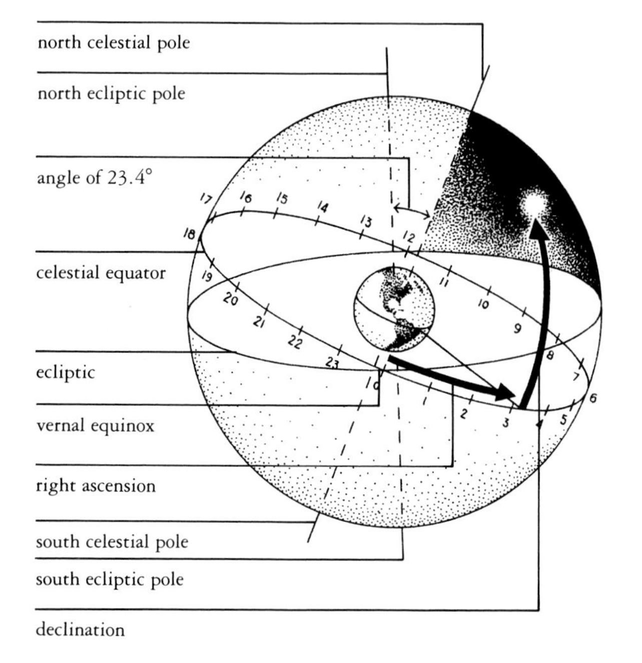

of arc; and one second (1^s) of right ascension equals 15 seconds ($15''$) of arc. The "zero point" of right ascension, celestial kin to Earth's Greenwich meridian of longitude, is the point at which the Sun (and thus the ecliptic) crosses the celestial equator on its way north each spring; this point is called the *vernal equinox* or the *first point of Aries*, because in ancient Greek times the Sun was in the constellation Aries on the first day of spring. The equatorial coordinate system is the one used most frequently to specify the positions of fixed celestial objects. You will see these coordinates plotted on the constellation charts. Catalogs of stars and of deep-sky objects specify the right ascensions and declinations of these objects at some particular time, or epoch. These coordinates change slowly because of precession. Remember that the directions "north" and "south" in the celestial sphere mean toward the north and south celestial poles, respectively. "East" always means toward your eastern horizon.

Angular Size

You can measure the approximate angular sizes of objects in the celestial sphere—or the distance between two objects—with your fist. Held at arm's length it is about 10° across at the knuckles. The tip of your index finger measures about 2°. Viewed from Earth the Sun and Moon are each ½° across (or 30′, 1,800″, or 2^m). If you read in the text for a constellation that an object is 4° north of another, you can use your fist to measure the distance.

Time and the Sky

As Earth turns, an observer at any particular location sees different parts of the celestial sphere in his or her line of sight. Every day we rotate 360° to the east; every hour, 15°; every 4 minutes, 1°. As seen from Earth, objects in the sky appear to move from east to west at this rate. For precise observations it is necessary to know the exact time at your location. Every night the stars rise 4 minutes earlier than they did the night before (this is because of Earth's movement along its orbit). Over the course of a month, about 30 days, stars will be rising 2 hours earlier (4 minutes/day $\times$ 30 days = 120 minutes). This is why you will see the same stars at 11:00 P.M. December 15 as you will see at 9:00 P.M. January 15. In the Sky Tours we have provided a range of times and dates for each seasonal sky chart, based on this principle.

Star Classification

Stars are assigned a spectral type and a luminosity class (explained in the essay "Stars" in the Introduction). These tables will help you decipher a star's classification and determine its type, color, and temperature.

Spectral Types

Type	Color	Temperature range (Kelvin)	Example
O	Blue	25,000–50,000°	δ Orionis
B	Blue	11,000–25,000°	Rigel (β Orionis)
A	Blue-white	7,500–11,000°	Sirius (α Canis majoris)
F	White	6,000–7,500°	Procyon (α Canis minoris)
G	Yellow-white	5,000–6,000°	The Sun
K	Orange	3,500–5,000°	Arcturus (α Boötis)
M	Red	3,000–3,500°	Antares (α Scorpii)

Luminosity Classes

Supergiants	Ia, Iab, Ib
Bright giant	II
Giant	III
Subgiant	IV
Main sequence	V
Sub dwarfs and dwarfs	VI, VII

The Brightest Stars

Name /Constellation	Distance from Earth	Spectral type	Apparent magnitude	Absolute magnitude
Sun	0.000016 ly	G2 V	−26.72	+4.85
Sirius (α) / Canis Major	8.8 ly	A1 V	−1.46	+1.42
Canopus (α) / Carina	74 ly	A9 II	−0.72	−2.5
α Centauri A / Centaurus	4.3 ly	G2 V	−0.01	+4.37
Arcturus (α) / Boötes	34 ly	K2 III	−0.04	+0.2
Vega (α) / Lyra	26 ly	A0 V	+0.03	+0.6
Capella (α) / Auriga	42 ly	G6 III	+0.08	−0.4
Rigel (β) / Orion	1,000 ly	B8 Ia	+0.12	−8.1
Procyon A (α) / Canis Minor	11.4 ly	F5 IV–V	+0.37	+2.64
Achernar (α) / Eridanus	69 ly	B3 V	+0.46	−1.3
Betelgeuse (α) / Orion	310 ly	M2 Iab	+0.5 variable	−7.2
Agena (β) / Centaurus	320? ly	B1 II	+0.6 variable	−4.4
Altair (α) / Aquila	17 ly	A7 IV–V	+0.77	+2.3
Aldebaran (α) / Taurus	60 ly	K5 III	+0.86 variable	−0.7
Spica (α) / Virgo	220 ly	B1 V	+0.91	−3.3
Antares (α) / Scorpius	325 ly	M1 Ib	+0.92 variable	−5
Pollux (β) / Gemini	36 ly	K0 III	+1.16	+1.0
Fomalhaut (α) / Piscis Austrinus	22 ly	A3 V	+1.19	+2.0
Deneb (α) / Cygnus	1,500 ly	A2 Ia	+1.26	−7.0
β Crucis / Crux	425 ly	B0 III	+1.28 variable	−4.6
Regulus (α) / Leo	69 ly	B7 V	+1.36	−0.7

Glossary

Alpha star
The Bayer designation of a star; usually the brightest star of a constellation. (Also α star.)

Ancient constellations
Forty-eight constellations originally identified in the second century by Ptolemy in his work *The Almagest.*

Apparent magnitude
The brightness of a star as perceived from Earth. A 1st-magnitude star is bright; the magnitudes of brighter stars are expressed in negative numbers. The dimmest stars visible to the human eye are of about 6th magnitude. A star's actual luminosity is called intrinsic or absolute magnitude.

Asterism
A separate identifiable shape within a constellation. The Big Dipper is an asterism within Ursa Major.

Bayer designation
A system of assigning Greek letters to stars, consisting of a lowercase Greek letter followed by the possessive form of the name of the constellation.

Black hole
Intense gravitational fields at the centers of some galaxies so powerful that even light cannot escape their pull.

Constellation
A particular region of the sky often enclosing a "figure" that was recognized long ago; every portion of the sky belongs to a constellation. See ancient constellations, modern constellations.

Critical density
When clouds of gas and dust in space reach a stage at which mutual gravity becomes strong enough to continue pulling the cloud together.

Deep-sky objects
Groups of stars (such as star clusters and galaxies) or nonstellar objects (such as nebulae) that exist beyond our solar system.

Ecliptic
The plane of Earth's orbit; the band of the sky through which, from our point of view, the Sun, Moon, and planets move.

Equinox
Two days of the year when day and night are of the same duration; the vernal equinox marks the beginning of Northern Hemisphere spring, and the autumnal equinox marks the beginning of fall.

Galaxies
Aggregates of gas, dust, and millions or billions of stars held together by mutual gravitational forces; classified as elliptical, spiral, barred spiral, irregular, and peculiar.

Galaxy cluster
A grouping of galaxies.

Index Catalog
Common listing of deep-sky objects, in the form of the letters "IC" followed by a number.

Light-year
The distance light travels in a year (abbreviated as "ly").

Local Group
Small cluster of galaxies to which our Milky Way belongs.

Luminosity class
A classification system that groups stars according to size; main sequence stars are luminosity class V, supergiants Ia, Iab, or Ib, white dwarfs VI or VII.

Main sequence stars
Stars that produce energy by converting hydrogen to helium (nuclear fusion).

Meridian
A line that runs from the exact north point on an observer's horizon up through the zenith and down to the southern horizon; also called the celestial meridian.

Messier list
Common listing of deep-sky objects (called Messier objects), in the form of the letter "M" followed by a number.

Modern constellations
Constellations identified in the 17th and 18th centuries; most are in southern skies.

Multiple stars
Stars that have one or more stellar companions; classified as optical, spectroscopic, and eclipsing binaries or multiples.

Nebulae
Clouds of gas (mostly hydrogen) and dust (mostly carbon and silicon) in space; classified as absorption, reflection, emission, planetary, or diffuse nebulae, or supernova remnants.

Neutron star
The core star that is left after a supernova explodes; a small star that rotates rapidly, sending out beams of light and radio waves; also known as a pulsating star or pulsar.

New General Catalog
Common listing of deep-sky objects, in the form of the letters "NGC" followed by a number.

Nuclear fusion
The process by which stars produce energy; the fusing of atomic nuclei (such as hydrogen) to form heavier nuclei (helium), yielding enormous quantities of energy.

Optical double or multiple
A pair or group of stars that appear close together along our line of sight but are physically unrelated.

Polestars
Stars that mark the north and south celestial poles; extensions of the terrestrial north and south poles into space, as determined by the direction of Earth's axis.

Precession
The slow gyration of Earth's axis; it takes about 26,000 years to complete one gyration.

Red giant
A star that has used up its hydrogen fuel; the star expands, its surface cools, and it sheds its outer atmosphere (which becomes a planetary nebula).

Spectral type
A system of classifying stars by the strengths and positions of absorption lines in their spectra, and thus by their color and temperature.

Star classification
A star's spectral type and luminosity class.

Star clusters
Groups of gravitationally bound stars similar to one another in age and composition that formed from the same interstellar cloud at about the same time; classified as open or globular clusters.

Supercluster
A grouping of galaxy clusters.

Supergiant
A very massive star that has used up its hydrogen fuel and begun to expand and cool.

Supernova
An exploding supergiant.

Supernova remnants
The nebulous debris left in space after a supernova explosion.

Variable stars
Single stars that vary in light output; classified as pulsating, eruptive, and rotating variables.

White dwarf
A very dense, hot star that remains after a red giant has ejected its outer envelope.

Zenith
The point directly overhead from an observer's point of view.

Zodiac
The band of sky extending above and below the ecliptic, containing 12 constellations.

Index

P

R

S

T

U

V

W

Z

Antique constellation charts (title page, pages 32–33, and 66–67) courtesy of *Le Dimore del Cielo,* Novecento Editrice (Italy). The constellations in these antique maps appear reversed, because before modern times astronomers plotted the stars as if from outside the universe, looking down at Earth.

This book was created by Chanticleer Press. All editorial inquiries should be addressed to:
Chanticleer Press
568 Broadway, #1005A
New York, NY 10012
(212) 941-1522

To purchase this book, or other National Audubon Society illustrated nature books, please contact:
Alfred A. Knopf, Inc.
201 East 50th Street
New York, NY 10022
(800) 733-3000

Chanticleer Press Staff
Publisher: Andrew Stewart
Managing Editor: Edie Locke
Art Director: Amanda Wilson
Production Manager: Susan Schoenfeld
Photo Editor: Giema Tsakuginow
Photo Assistant: Consuelo Tiffany Lee
Publishing Assistant: Alicia Mills
Text Editor: Amy K. Hughes
Copyeditor: Patricia Fogarty
Natural Science Consultant: Richard Keen
Diagrammatic drawings: Ed Lam
Constellation illustrations and cover art: Acme Design Company
Original series design by Massimo Vignelli.

Founding Publisher: Paul Steiner